Getting into Print

OTHER TITLES FROM CHAPMAN & HALL AND E & FN SPON

Effective Speaking
Communicating in speech
C. Turk

Effective Writing
Improving scientific, technical and business communication
2nd edition
C. Turk and J. Kirkham

Good Style
Writing for science and technology
J. Kirkham

Scientists Must Write
A guide for better writing for scientists, engineers and students
R. Barrass

Studying for Science
A guide to information, communication and study techniques
E.B. White

Write in Style
A guide to good English
R. Palmer

Writing Successfully in Science
M. O'Connor

Brain Train: studying for success
2nd edition
R. Palmer

Study!
A guide to effective study, revision and examination techniques
R. Barrass

For more information on these and other titles please contact:
The Promotion Department, E & FN Spon, 2–6 Boundary Row, London SE1 8HN.
Telephone 071 865 0066.

Getting into Print

A guide for scientists and technologists

Peter Sprent

Emeritus Professor of Statistics
University of Dundee
UK

E & FN SPON
An Imprint of Chapman & Hall
London · Glasgow · Weinheim · New York · Tokyo · Melbourne · Madras

**Published by E & FN Spon, an imprint of
Chapman & Hall, 2 – 6 Boundary Row, London SE1 8HN**

Chapman & Hall, 2 – 6 Boundary Row, London SE1 8HN, UK

Blackie Academic & Professional, Wester Cleddens Road, Bishopbriggs,
Glasgow G64 2NZ, UK

Chapman & Hall GmbH, Pappelallee 3, 69469 Weinheim, Germany

Chapman & Hall USA, One Penn Plaza, 41st Floor, New York NY 10119, USA

Chapman & Hall Japan, ITP-Japan, Kyowa Building, 3F, 2-2-1 Hirakawacho,
Chiyoda-ku, Tokyo 102, Japan

Chapman & Hall Australia, Thomas Nelson Australia, 102 Dodds Street,
South Melbourne, Victoria 3205, Australia

Chapman & Hall India, R. Seshadri, 32 Second Main Road, CIT East,
Madras 600 035, India

First edition 1995

© 1995 Peter Sprent

Printed in Great Britain by Page Bros, Norwich

ISBN 0 419 19220 4

A catalogue record for this book is available from the British Library

∞ Printed on permanent acid-free text paper, manufactured in accordance
with ANSI/NISO Z39.48-1992 and ANSI/NISO Z39.48-1984 (Permanence
of Paper).

Contents

Preface

Scientists and technologists have to communicate. Getting into print is often the key to a successful career. Competent writing may bring direct financial reward, but the bonus is often better career prospects. Those who do not write well seldom scale the professional peaks.

It is not intrinsically hard to give a clear account of facts and ideas; this does not call for great literary skill. You need only master some basic rules and apply them in a common-sense way. This book is about those basics and the way they must be fine-tuned for each kind of writing. The objectivity of good scientific and technical writing often curbs the individuality that is an important ingredient of much creative literature.

Many *how-to-write* books concentrate on a particular kind of writing — crime fiction, romantic novels, historical novels, biographies, scientific and technical papers, instruction manuals, theses and dissertations — the list is long. Some *how-to-write* books are highly regarded, but even among the good ones, a guide that some authors find excellent will disappoint others because it is not tailored to their needs. Writing a research monograph calls for an approach that is different from that for writing detective fiction. A technical report is compiled in a way that would be inappropriate for a school textbook.

Some types of writing about science and technology — what I call the *have-to-do* tasks — are well covered, but there are few books that look at scientific and technical writing as a whole. That is my excuse for adding to the long list of *how-to-write* books. Bridging this gap is important because, despite that need to communicate, many scientists or technologists receive little or no formal training in some of the relevant skills.

In compiling this book I look upon scientific and technical writing as relevant not only to the physical and life sciences, to medicine, to industrial research and development but also to the social sciences and to manufacturing, business and commerce, for much of what I say applies to writing in all these areas.

There are many interlinking threads, so I hope all readers will at least skim through every chapter. Please do this even if some do not appear immediately relevant to the sort of writing you want, or have, to do. I say a lot about writing books because that is a task that brings together all the important writing skills. Clarity, correct spelling and punctuation, and basic good grammar are equally necessary for reports, manual writing, and preparing material of a technical or popular scientific nature for the media. However, writing books has a completeness, compared to which other writing tasks are something of a literary *coitus interruptus*. I do not mean that such writing is unimportant, undemanding or trivial. Indeed, to carry the analogy further, that other writing often requires more discipline than going the whole way. Whatever you write should match the task at hand — in content and in style and presentation.

The core chapters are chapters 1, 2, 3 and 6. If you want, or have, to write books you should also read carefully chapters 4, 5, 7 and relevant parts of Chapter 10; others may be content to read these chapters selectively. Chapter 8 is mainly about work-related scientific and technical writing other than books, and Chapter 9 is about writing for more 'popular' markets. These divisions are not, and are not meant to be, watertight. Again, let me stress that many points I make in relation to a specific form of writing are relevant to other aspects of written, and often also to verbal, communication.

Peter Sprent

Wormit, Fife
Scotland

About the author

Peter Sprent is Emeritus Professor of Statistics at the University of Dundee, Scotland. He has written nine books, ranging from a research monograph to student textbooks and popular works on mathematics and statistics. He has also written many expository and research papers, prepared and presented scientific material for radio and television and has been editor of several leading international statistics journals.

1

Communicating

1.1 LEARNING ABOUT WRITING

The last thing one knows in constructing a work is what to put first.

The famous French mathematician and theologian, Blaise Pascal (1623–62) recorded this simple truth. It is one reason why so many scientists and technologists find it hard to write.

There are many *how-to-write* books to help potential authors decide what to put first, and how to go about the rest of their task. Most cover one or more of the following aspects of the art of communicating.

- **Style** — punctuation, spelling, grammar, choosing words and structuring them to capture and hold readers' attention and to achieve an objective (which may be to inform, teach, explain, entertain, stimulate emotions or provoke thought).
- **Specific categories of writing** — including general fiction, crime fiction, romance, biography, journalism, radio plays, poetry, film scenarios, newspaper and magazine articles, school textbooks, research reports, instruction manuals.
- **Getting published** — finding a publisher, publishers' requirements, contracts, proof-reading, marketing, selling to magazines.

In this *how-to-write* book I try to smooth your path to, and through, the publishing process if you write about science or technology. A sound style is essential, so I say a lot about that. John Kirkman, in *Good Style — Writing for Science and Technology* (FRI), tells you even more; he writes specifically for those who prepare reports and instruction manuals, or papers for technical, learned or professional journals. That (FRI) after the title means that details of the book appear in section I of the **Further reading** list on p. 204. I use an abbreviation FR*n*, where *n* is either I, II, or III, often; it refers you to section *n* of the reading list for more information. Michael Legat, in *The Nuts and Bolts of Writing* (FRI) deals with style in general terms; he also gives useful tips for all writers hoping to get published, but he does not cover matters peculiar to science or technology. At the opposite extreme there are some very specialized books; for example, Nicholas Higham's *Writing for the Mathematical Sciences* (FRIII) puts emphasis on style and techniques specially relevant in one narrow field. To keep this book to a reasonable length and price, I shall often refer to, rather than repeat, advice that is readily available elsewhere.

1.2   WRITING ABOUT SCIENCE AND TECHNOLOGY

Scientific and technical writing splits broadly into *have-to-do* and *want-to-do* writing, but there is no clear divide between the two. Research workers must report new findings; engineers, industrial chemists, computer programmers and others, especially if they work in an industrial or commercial environment, often have to prepare reports for management and for colleagues, and write articles for trade or professional journals as well as memos and letters that convey technical information. Thousands of people, many technically qualified, produce instruction manuals for equipment or for computer software. These are usually *have-to-do* tasks; the people who do them, *have-to-do* writers.

If it is part of your job to write a textbook or a research monograph you are also a *have-to-do* writer. If it is a labour of love, or you are inspired by an urge, you are in the *want-to-do* category. Authors of books for background reading by students or by school pupils, as well as those who write about their own discipline for scientists or technologists working in other fields, are usually, but not always, *want-to-do* writers. So are most of those — other than professional journalists — who popularize science, engineering, medicine or technology in books for the general reader, or in newspaper and magazine articles, or on radio and TV.

Many scientists and engineers are trained — formally or informally, extensively or incidentally — to prepare reports; this may extend to writing scientific or technical papers for academic, professional or trade journals. Some learn, often informally, how to write theses or dissertations. Handbooks, some catering for the British market, some for the American, and some for both, that cover these kinds of *have-to-do* writing are listed in FRI; many are excellent general guides to that elusive quality *style*. Advice about *want-to-do* scientific and technical writing is less readily available.

1.3 STYLE

A simplistic dictionary-type definition of *style* is 'the manner of writing'. In 1720, Jonathan Swift, in a letter to a young clergyman, equated *style* to a goal of *good style*:

Proper words in proper places, make the true definition of a style.

Make **proper words in proper places** your guiding principle. Finding the right words and putting them in the right places tests any author's skill. Books by Turk and Kirkman, by Kirkman, by Blake and Bly, by Turabian, by Rubens and by Day — all listed in FRI — are but a few of the many that tell you a lot about this aspect of style, especially for the classes of *have-to-do* writing each covers. Read at least one of these, also that by O'Connor (FRI), and you will understand the nuts and bolts of writing reports, papers describing research or technical innovation, theses and dissertations. These experts say a lot about style that extends to all good writing; but research or technical reports also impose restrictions that may, and should be, relaxed when you are writing a textbook, a book for background reading, one intended to expound your own discipline to workers in other fields, or to provide a 'popular' account of an aspect of science or technology for interested nonscientists.

I interpret style to cover more than Swift implies, and I have invented the words macrostyle and microstyle.

- **Macrostyle** is about general structure. For a book, that includes logical ordering of material, division into chapters, and usually sections, sometimes subsections. It embraces also policy decisions about whether to include examples, exercises, graphs, diagrams, tables, illustrations, further reading lists, a bibliography or references, appendices, glossaries, an index, or other features such as computer program listings. In essence, macrostyle has to do with large groups of words, or with blocks of information.
- **Microstyle** embodies Swift's notion of style, beginning with choosing proper words, then using these as building blocks for sentences and paragraphs.

Microstyle merges with macrostyle when paragraphs are assembled to form sections or chapters. Individual tables and diagrams have an internal structure of their own; I think of this as their *microstyle*.

1.4 FACING THE HURDLES

Writing is hard work, but I enjoy it. I was lucky; at school I was taught basic writing skills. I was drilled in grammar, spelling and composition as part of the core curriculum, although that now-fashionable phrase had yet to be invented. That was just a start; like all writers, I have learnt as I work. Mistakes have often provided salutary lessons.

For several decades many schools have given pupils only minimal training in grammar and spelling; this has not helped potential authors. Expounding their work comes naturally to a lucky few, but most people can do it only after a lot of practice. Even those who find it easy to pass information to their peers, or to close colleagues, may have problems describing their work to students, or in explaining it clearly to a layperson.

1.5 INSPIRATION

What inspires *want-to-do* writers? You may get a sudden impulse to write, but more often the slow gestation of an idea will convince you that you have something worth recording. If it calls for a book, this may take months, even years, to complete. Your first ideas will be vague. You may hope, even expect, that they will become clearer as you proceed. They might; but if not, your book may never be finished. You will certainly save time if you preplan.

There is no master blueprint for preplanning. What I do may not suit you. I can only make suggestions and give tips that have helped me, or that I know have helped others. If I implore you to follow a particular route, it is because many writers have found it useful; but apart from a few *musts* that apply to all authors, how you write is a personal matter.

Having decided to write, think carefully about what you want to produce. A book? A paper? A magazine article? What kind of reader do you hope to attract? This helps you shape your ideas about precisely what to write. Try to visualize, indeed define, a **target readership**, especially if you are writing a book. There is a feedback mechanism here; this target will influence the way you write.

1.6 DIRECTING EFFORT AND MEASURING SUCCESS

Successful books about science or technology — indeed about most things — are almost always written for a predefined group of readers; these may be peers in your own or a related discipline, they may be students, school pupils, the enquiring layperson, or a wider but interested public. Your precise target is generally a subset of one of these broad groups. It might be undergraduate engineers, or school and college physics students, or the amateur with an interest in geology.

For a chosen group you could write more than one kind of book; for students it might be a textbook, one for background reading, a critical account of recent developments in your subject, a guide for a service element in some course, e.g. a book on chemistry for concrete technologists, one on mathematics for biologists, or an introduction to parasitology for agronomists or for zoologists.

You will want your book to be a success. All writers do. Measures of success range from academic or professional recognition to financial gain. Your idea of success will influence the way you write. Acclaim by your peers does not always imply financial reward or vice versa. Realistically, especially if writing is only a part-time activity, you are unlikely to make a fortune by writing books about science and technology. I am not saying you won't, or you can't, but to do so is rare.

Your may decide to write a monograph, or treatise, because results in your line of research are available only in papers in several journals, and you see a need to collate this work. If your subject is the nerve structure of *Anaspidacea*, your readers will probably be fewer than a hundred experts worldwide. Write about the potential role of genetic engineering in preventing cancer, and do it well, if there is no competing volume your expected readership will be quite large. For a specialist monograph, acclaim by fellow experts, or from the wider scientific community, is the usual mark of success; significant financial gain from the writing *per se* is unlikely, though scholarship evident in the work may enhance your career prospects.

There is a steady demand for textbooks and for background reading material for students. This is a competitive market. Your target readership may be postgraduates, undergraduates at levels from elementary to advanced, or secondary, even primary, school pupils. Many textbooks are 'tailored' to specific examination syllabuses. A successful textbook is one that is widely adopted for class use, often after receiving good reviews; this does not guarantee big financial gains for author or publisher, though many textbooks are sound commercial propositions. There is also a market for *self-tuition* and *do-it-yourself* books, the latter usually having a technical flavour.

You may have an evangelical urge to introduce your speciality to other scientists, or to alert the interested amateur. Books for the latter sometimes aim to entertain as well as inform. For this kind of writing, favourable reviews and good sales are measures of success; the former do not guarantee the latter, but poor reviews are commercially unhelpful and ego-deflating.

New instruction manuals are needed constantly. A problem facing their writers — and one that is often neglected — is that users of the equipment, computer software, or whatever is the subject of the manual, may have relevant background knowledge ranging from nil to advanced. A good instruction manual must be intelligible to all users without being patronizing to those who already have good background knowledge. Success for manual writing is hard to measure directly; a lack of complaints, few requests for explanations of obscure passages, or grumbles about difficulty in following the instructions, are good omens. They imply that the manual is probably doing its intended job without creating serious problems. Remember, though, that many people are reluctant to complain, seeing that as an admission of inadequacy. Writing a manual is often part of one's job, a *have-to-do* task, so there may be no equivalent to the 'royalty' payment that applies to general books. If you are commissioned to write a manual the fee is often a matter for negotiation.

Apart from books or manuals you might also want — even be asked — to write an article for a newspaper, or a widely read journal like *The New Scientist* or *Scientific American*, or for a periodical aimed at the enthusiast on computing, flying, motoring or some other technical subject. Payment, often negotiable, may vary from zero to £100 or more per thousand words. If you work in industry or commerce you may have to write 'nonspecialist' technical reports or expository articles for internal circulation in your company, or for a trade or an industrial association. If this is part of your job don't expect extra pay, but doing it well may help your promotion prospects.

Finally, you may want to write scripts about your work for radio or television. Such work is often well paid, but these are competitive markets and often the first problem is to break into them.

1.7 SOME GENERAL PRINCIPLES

I consider both *macrostyle* and *microstyle* in later chapters, but there are general principles to keep in mind from the word 'go'. Be clear. Distinguish between facts and opinions, between hypotheses and truths. This applies to all kinds of scientific and technical writing.

Each discipline has its own vocabulary and jargon. Distinctive meanings are often assigned to common words; *significance* means something special to a statistician. A technical term may not have the same meaning in all disciplines. A mathematician calls an entity in Euclidean space that has both magnitude and direction a *vector*; to a plant pathologist a *vector* is an organism that carries a disease from plant to plant. Jargon, often in the form of abbreviations or acronyms, is common, especially in rapidly advancing fields like computing and information technology. Even if you are not a computer buff you may have met *RAM* (random access memory), *ROM* (read-only memory), *wysiwyg* (what you see is what you get) and *I/O* (input/output). New jargon is constantly being invented, and is useful in the right context. I invented *macrostyle* and *microstyle* in section 1.3.

Most disciplines have a corpus of accepted terminology and spelling. This may run contrary to your instincts; in the UK chemists now prefer *sulfate* to *sulphate*, the former long preferred by everybody in the USA. This shift by their chemical colleagues does not please all UK biologists or agronomists, although it is the preferred option for their French counterparts. You might, especially if you are not a chemist, get away with *sulphate*, but you should not depart from most conventions without good reasons. For example, because SI units (the international system) were adopted to remove difficulties in comparing results expressed in a variety of units, you need a good reason for not using them. These reasons exist, especially if you are writing for the US technical market, or for certain 'application' groups such as the aviation industry (section 3.4.2). Incidentally, the term *SI units* is itself scientific jargon!

Capturing and holding readers' interest is an elusive facet of success. Clarity and crispness are essential; so are pace and amount of detail. Go too fast and you will frustrate readers; move too slowly, or put in too much detail, and you will bore or irritate. Generalities hold most people's interest only for a limited time. Relevant generalities are important, but they are less interesting than specifics.

1.8 GETTING BOOKS PUBLISHED

If you are writing a book, when you first set pen to paper — or more likely switch on your word-processor — publication may seem a distant goal. You may miss that goal unless, at this early stage, you clarify your aims and define (provisionally at any rate) a target readership. An early step is to prepare a proposal for a publisher (section 4.1); but read chapters 2, 3 and 4, and perhaps refer to any relevant sections in chapters 6 and 7, before you do this.

Some writers complete the full draft of a book before seeking a publisher. That may be the best, or the only, way to sell a first novel, but there is little to commend it for science or technology. It has wasted *your* time if you cannot find a publisher, and the days are gone when publishers would accept almost anything from a known expert. What is more important, a publisher's commissioning editor, given outline contents and a couple of specimen chapters, will often enter into a dialogue that will influence the way your book develops and may save you hours of unnecessary work. That happened with this book.

Follow the guidelines in sections 4.1 and 4.2 when looking for a publisher. Approach only those who *currently* market the sort of book you are writing. Study up-to-date publishers' catalogues, or if these are not available, look at recent books in an academic, technical or public library to find out who publishes what. The annual *Writers' and Artists' Yearbook* (FRII) and *The Writer's Handbook* (FRII) list major UK and American publishers indicating the type of work each handles; the former also lists other overseas English-language publishers. Unfortunately, but understandably, both give only brief descriptions that are plagued by vague terms like *general nonfiction* or *technical and academic*; but checking will at least warn you not to inflict your monograph on *Hormone Secretions by Vampire Bats*, no matter how these may stimulate those mammals' sex lives, to Mills & Boon. Use current editions of these guides, because publishers' requirements frequently change. For the American market, *The Writers' Market* (FRII) and the US *The Writer's Handbook* (FRII) — no relation to its UK namesake — are the corresponding guides. Personal contacts are also useful. A colleague who knows an appropriate publisher or commissioning editor may give you an introduction that could save you weeks, even months, of frustration.

When you have a contract, complete your manuscript by the agreed date, provide information for marketing when asked, correct proofs and prepare an index by the date requested. Authors who are slap-happy about these matters are their own worst enemies.

Modern computer technology makes self-publishing (section 4.4) more attractive than it was a decade ago, but there are pitfalls. However, if your proposed book has a small potential readership self-publishing may be virtually the only way of getting it into print.

1.9   THE LAYOUT OF THIS BOOK

Chapters 2 and 3 deal with macrostyle and microstyle. Steps to follow when seeking a publisher, and when you have found one, ways to keep the author and publisher relationship on an even keel, are covered in chapters 4 and 5.

Chapter 6 looks at devices and tools at the disposal of scientific and technical writers, while Chapter 7 is about writing books of various kinds. Chapter 8 is devoted to other aspects of writing about science and technology for one's peers, or for those who need to know about scientific and technical concepts or equipment, including the preparation of manuals. Chapter 9 is about semi-technical and popular writing for the media. Chapter 10 deals with miscellaneous matters including how to liaise and work with copy-editors; some potential problems with joint authorship and with contributed chapters to multi-author volumes; contributing to reference volumes such as subject dictionaries; special considerations for authors writing in English when this is not their first language; sources of help for authors; payments for library loans and copying. Appendices give guidance on copyright and other legal pitfalls; some hints for preparing CRC are illustrated by examples from this book; there is also a glossary of publishers' and printers' terms and jargon that authors often meet.

2

Macrostyle

2.1 FIRST STEPS

Decide the type of book you want to write and define your target readership; make decisions about style. Start with macrostyle, i.e. determining *proper places* for relatively large groups of *proper words* — sections, chapters and sometimes parts; also make broad policy decisions about whether, and how, to use devices like:

- photographs;
- diagrams;
- graphs;
- tables;

and, depending on your subject, whether to include:

- examples;
- exercises;
- descriptions of laboratory experiments;
- computer program listings.

2.2 PLANNING YOUR WRITING

2.2.1 Defining a target readership

Think carefully about what kind of reader you expect, or hope, to attract — your target readership — before making key decisions about style. Do this because your style should match readers' expectations. Target readership and appropriate style both depend upon the type of book. A specialized monograph aimed at research workers studying the role of cell division inhibitors in treating cancer, or one explaining to theoretical physicists the latest developments in the special theory of relativity, will have a target readership that is almost self-defining; namely experts in the relevant fields. Your style should take account of that expertise; for example, you may use, without explanation, concepts or jargon that will be familiar to these specialists.

If you plan a school textbook to cover the English 'A'-level physics syllabus or an American, Canadian, Indian, Kenyan, Australian or any other high school or college physics syllabus, the target looks to be all pupils studying that syllabus. This may be too ambitious, because there are many established texts for most school syllabuses. Is there a need, or a market, for a new one? Ask that question if you want to write a textbook in any highly-competitive field. To answer it sensibly, you need to know what texts are currently in use and how popular each is with teachers and pupils. Are there serious gaps in their syllabus coverage? If there are no clear deficiencies, publishers may be unenthusiastic about a new book. If current textbooks all ignore, or deal inadequately with, some aspect of the syllabus, there may be a market for a book that overcomes that deficiency. But be warned; scant coverage of part of a syllabus may reflect a general lack of teacher, and pupil, interest in a topic.

Writing school texts is a specialized field; teaching experience at the appropriate level is virtually a prerequisite for authors who enter it.

Postgraduate, or undergraduate, texts for college and university students again appear to have a defined target readership, but heterogeneity may be a problem. The content of undergraduate courses, and the nature of post-graduate instruction in any subject, vary between institutions. If you are writing for, say, postgraduate biochemistry students, you might aim your book at only a specific subset of these (e.g. those interested in cell biochemistry), or you could target a broad spectrum of biochemists. Whatever your decision, think carefully about what you may reasonably assume readers will already know, and how much explanation of new concepts will be needed.

A glossary of specialist terms that are likely to be familiar to some, but not to all, potential readers, often helps a heterogeneous readership. Concepts or techniques that may be familiar only to some readers might be described in appendices. The smooth flow of a book is interrupted for those who need to refer to these appendices, but this is a lesser evil than annoying advanced readers by almost forcing them to plough through material that is already familiar. Footnotes may also help in these situations, but use them sparingly.

In section 2.2.2 I make the case for attractive layout; footnotes spoil this; they also give problems to printers, although modern technology has reduced these.   Whether to use glossaries, appendices, or footnotes are clearly decisions about *macrostyle*.

Getting the presentation level right is hard when you write for scientists or technologists who are not specialists in your subject, even harder if you write for the interested layperson. If you are tempted to introduce a new concept, or use jargon, ask yourself if it is necessary or helpful; introduce it only if the answer is a justifiable 'yes'. Beware of self-interest. You may enjoy — though I would find it hard to believe — calculating the volume of a bucket, given its height and basal and top diameters, but do not quote a formula to do this if all that your readers need to know is that most household buckets have a capacity of about 10 litres.

Science and technology are becoming more interdisciplinary. A common mistake made by specialists in one subject, when trying to communicate with workers in another, is to refer to a concept or to use terminology, which, although correct, is unhelpful. Here is an example from my own subject, statistics. A statistical method often used in biology, chemistry, engineering, medicine, economics, industrial production planning and marketing is called regression analysis.   Major statistical programs, and many spreadsheet packages include regression analysis. There are several situations where these programs fail because of what a statistician calls a *matrix singularity*. One cause of this difficulty is not having enough data. In that event a sensible statistician tells the client that there are not enough data. Clients who are, for example, biologists or marketing managers, will understand this, and come to a rational decision either to collect more data or to abandon the regression analysis. A statistician who tries to blind the client with science by saying 'You have a matrix singularity', may leave the victim wondering whether to seek medical advice, or just abandon hope of coming to grips with statistical methods.

Instruction manuals for equipment are often criticized because they use unexplained jargon. These manuals are often written by technical experts, who sometimes ignore, or underestimate, the gap between their own technical knowledge and that of likely users.

2.2.2 Attractive layout

Authors and publishers share responsibility for the layout of a book. The publisher's role used to dominate, but for scientific and technical books the situation is changing. Firstly, because many of these books involve formulae, equations, tables, technical diagrams, etc., and the author may sometimes know how best to present these. Secondly, modern computer technology makes it feasible for authors, with their publisher's agreement, to typeset, or partially typeset, their work; copy may be supplied on disc with agreed typesetting codes inserted, or the manuscript may be prepared as camera ready copy (CRC). I elaborate on these possibilities in sections 5.2.2 and 5.2.3, concentrating here on the author's responsibility for basic layout.

Detailed layout is usually finalized when a sub-editor, copy-editor or designer marks up a manuscript for typesetting, but decisions about whether a book splits naturally into parts (e.g. Part I and Part II), about chapter content, and whether to split chapters into sections and sections into subsections, must be made by the author at an early stage.

Scientific or technical books often have text within chapters broken into headed, and sometimes numbered, sections, often with displayed formulae and many tables, diagrams or illustrations. Material, such as descriptions of experiments, examples and exercises, may be indented or set in font or point sizes different from the main text. Although it is not the primary reason for using them, these features, used properly, make the layout more attractive.

Such tools are seldom available to writers of fiction or of nonfiction like biography or memoirs, where free-flowing prose is broken only into chapters, paragraphs and sentences. A novelist who aims to entertain (e.g. one who writes romantic novels or crime fiction) relies on choice of a plot and skill as a story-teller to hold readers' attention. A serious novelist wanting to deliver a message chooses an appropriate plot and crafts words into prose designed to trigger chosen responses from readers, perhaps inducing a sense of shock, bewilderment, insecurity, or one of euphoria. All this requires great skill in the use of language and often an understanding of psychology.

By contrast, writers of most scientific or technical 'literature' are communicating facts, postulating theories, making critical assessments. These seldom entertain or appeal to emotions and, like taking unpleasant medicine, their assimilation calls for effort. Sweetening the pill may encourage the patient, i.e. the reader, to swallow the medicine. Attractive layout is a potential sweetener, but it needs a health warning. Just as a medicine only cures if it is appropriate, good layout will not redeem a bad book, but poor layout may lose potential readers for otherwise sound work.

2.2.3 Sections and section headings

Headed sections and subsections discipline writers and help readers. Division into sections forces writers to think about logical order. Informative headings assist a reader looking for specific items. The structure into sections and subsections, if the latter are also used, will depend upon the subject and depth of treatment.

If your book is about the geology of the British Isles and targeted at amateur geologists wanting information about rock features in places they visit, you might devote a chapter to each of a number of regions that are, geologically, reasonably homogeneous. It would, incidentally, be sensible to include a map showing these regions. One region might be that geologist's paradise, the Western Highlands of Scotland. You might divide each chapter into sections, a section describing the geology of a recognized sub-region within the overall area. For the Western Highlands, the sub-regions might correspond to geographical or administrative divisions, e.g. Invernessshire, Wester Ross, Sutherland, the Inner Hebrides, the Outer Hebrides . . .. You might prefer to sub-divide on geological criteria, e.g. rock types; each section describing a different rock type found in one or more places in the Western Highlands.

The target readership should guide your choice. If it is tourists with broad geological interests who are likely to visit only one or two specific areas, the breakdown by administrative districts would help. If a majority of readers are likely to be interested only in one or two particular kinds of rock, a breakdown by rock type would be more useful. You could cater for both — doing this in one of the following two ways.

- Devote sections to each administrative division in the Western Highlands and subsections to each rock type found within that division.
- Devote sections to each major rock type and subsections to each administrative division where that type occurs.

Visualizing target readers' reactions will help you choose between such alternatives. Avoid a breakdown to sub-subsections unless there are pressing logical reasons; these over-fragment most writing.

A heading should tell readers what a section or subsection is about. A sociologist writing about social security payments for the unemployed might devote a section to special considerations for the disabled. Suitable headings would be **Disabled recipients** or **Disabled recipients — special factors**. Avoid vague generalities like **A special case.**

Section and subsection headings may be distinguished by different fonts or typefaces, so make it clear on your manuscript whether a heading refers to a section or to a subsection (often called *primary* and *secondary*, or *A* and *B* level, headings). If this is not clear from a numbering system, use different indentations for each, or pencilled marginal annotations (*A* for section, *B* for subsection, heading). Publishers often tell their authors how they like this to be done.

In this book I number sections so that I can refer to them easily from other parts of the book. Sections are numbered within chapters and subsections within sections, e.g. section 4.1 refers to section 1 in Chapter 4; section 3.2.4 refers to subsection 4 of section 2 in Chapter 3. Cross-referencing to sections or subsections, rather than to pages, saves work at the proof correction stage; for, unless one is preparing CRC, page numbers are not known before the type is set. If page referencing is needed, enter a cross-reference as 'p. 000' in the manuscript. Insert actual page numbers at proof correction.

I do not always use subsections. Section 2.1 has no subsections, but section 2.2 (this section) has; you are reading section 2.2.3. i.e. subsection 3. Sometimes (e.g. section 3.2) I put 'floating text' before the first numbered subsection (section 3.2.1). Some writers and publishers consider this untidy, but it is logical if a section starts with general remarks that apply to all later subsections. For example, if section 5.2 of a book on the environment deals with human factors causing erosion on hills and is headed:

5.2 HUMAN FACTORS

you may want first to make general remarks about human influences. These might be put at the start of section 5.2 and be followed by subsections devoted to particular causes, e.g.

5.2.1 **Erosion by walkers**
5.2.2 **Erosion by mountain bikes**
5.2.3 **Erosion by four-wheel-drive vehicles**

etc. If you want to avoid, or your publisher does not like, 'floating' text at the start of a section begin with a subsection '5.2.1 **Human influences on erosion**' and renumber the later subsections '5.2.2 **Erosion by walkers**', etc.

2.2.4 Tables, graphs, diagrams and illustrations

Whether, and how best to use tables, graphs, diagrams, photographs and so on depends on the nature of a book and the target readership. I make general remarks here; recommendations about particular types of tables, graphs, diagrams, etc. are given in Chapter 6.

Graphs and tables are often complementary ways of presenting information. Which is the more useful depends upon the nature of data, what they are being used to illustrate, and how target readers are likely to react. Bankers, accountants, financiers, mathematicians, engineers and others used to handling numbers may prefer a table to a graph, yet be reasonably happy with either. Those less used to dealing with numbers — many social scientists, biologists, doctors, managers, politicians, journalists — may prefer a good graph. Appropriate graphs have the edge when only general impressions are needed. Tables are better for showing subtle detail; useful also if readers may want to combine information or highlight specific aspects of data. Usually, either graphs or tables are better than data embedded in text.

I give a simple example, based on information given by Rahn (1993) in a technical report that discusses questions about integrating the East and West German railway systems. Here is the data embedded in text (my version, not that in the original paper):

> The West German railway system is the *Deutsche Bundesbahn* (DB) which, in 1991 had approximately 242 000 employees; the East German system, the *Deutsche Reichsbahn* (DR) had approximately 208 000 employees, making the joint total approximately 450 000. The DB had a network of 27 000 km, twice that of the DR network. Passenger traffic totalled 45 600 million passenger km on DB and 10 300 on DR. Freight totals were 62 100 million tonne km for DB, but only 18 600 million for DR.

Unless you have a good head for figures (a vague and imprecise catch-cry), the above information is hard to digest. It gives a general impression that the East German system, the DR, is the smaller with fewer staff, a shorter network and less passenger and freight traffic. Did you reach any conclusion about the relative efficiency of operation? These raw data are not useful for assessing efficiency without further calculations. What to do becomes clearer from a table. Incidentally, the measures of passenger and freight traffic, i.e. *passenger km* and *tonne km*, are both commonly used in transport statistics. The former represents, in essence, the sum of the distances, in km, that each passenger is carried, and the latter is obtained by adding, for each load of freight, the product of its weight in tonnes multiplied by the distance, in km, it is carried (or equivalently, the sum of the distances, in km, that each tonne is carried).

Table 2.1 is based on the information above. The table makes it easier to compare the two systems because comparative figures are arranged side by side in adjacent columns. It is a *raw data* table that provides a starting point for further analyses that may need derived tables.

What tables are relevant depends on how information is to be used. In this example an efficiency comparison based on *unit of track length* (i.e. 1 km) may be appropriate, comparing for each system the numbers of employees per km of track, passenger loads per km of track, and freight loads per km of track.

Table 2.1 Comparable data for the two German railway networks, 1991

	Deutsche Bundesbahn (DB)	Deutsche Reichsbahn (DR)
Network (km)	27 000	13 500
No. of employees	242 000	208 000
Passenger km (million)	45 600	10 300
Freight ton km (million)	62 100	18 600

This information is easily calculated from Table 2.1. For example, for the DB network the number of employees per km of track is 242 000/27 000= 8.96 and for the DR network it is 208 000/13 500=15.41. Similar calculations for passenger and freight performance per km of track lead to the results in Table 2.2.

Table 2.2 Relative performances per km track for German railway systems, 1991

	Deutsche Bundesbahn (DB)	Deutsche Reichsbahn (DR)
Employees km^{-1}	8.96	15.41
*Passenger load km^{-1}	1.69	0.76
*Freight load km^{-1}	2.30	1.38

*Passenger load and freight load are measured in million passenger km and million tonne km per annum respectively.

The footnote in Table 2.2 explains how passenger and freight loads are measured. I advised against footnotes in general text, but they are often appropriate in tables because a table should be self-contained and intelligible without having to refer to the main text. However, interpretation may, and usually will, be given in the text. In this example the text might draw attention to the fact that *per km of track*, DR has almost twice the staffing levels of DB, whereas both passenger and freight loads per km of track for DR are of the order of one-half those of DB. An account of these findings might also include possible explanations for these differences. For example, lower staff levels might reflect highly automated signalling, more modern equipment, and more efficient ticket issuing facilities for one system, or different employment policies for each system. Different loads may reflect social, geographic and economic factors.

To indicate one use of graphs, I give two simple examples using the data in Table 2.3. Many types of graph exist — histograms, bar charts, pie-charts, flow charts, etc. — and some are described in Chapter 6.

Table 2.3 Numbers of wet days and annual rainfall, 1978, for nine locations in West Scotland

Station	No. wet days (x)	Annual rainfall (mm) (y)
Benmore	200	2286
Brodick Castle	188	1851
Clatteringshawe	191	1874
Glasgow airport	162	1074
Millport	178	1095
Rothesay	186	1398
Sloy	196	2454
Stirling	145	1003
Threave	144	1104

Source: *Meteorological Office Monthly Weather Summaries*, HMSO, 1978.

From Table 2.3 it is easy to see that the numbers of rainy days (x) range from 144 to 200 and the annual rainfall (y) from 1003 to 2454 mm. I have chosen the x, y scales in Figure 2.1 so that these ranges 'fill' most of the x, y axes. This means that the origin is not at the intersection of these axial lines; I indicate this by the clear breaks in the axes near the bottom left of the figure. For clarity, always use a device like this if the origin is not at the intersection of the axes as drawn.

A common fault with graphs and diagrams is inadequate labelling: e.g. in data plots always indicate clearly on the axes what is being measured and what the units are. The point with x, y values corresponding to each of the nine data points is denoted by a closed circle (●) in Figure 2.1. Each axis is labelled and the diagram has an informative caption. This simple data graph shows that not only does annual rainfall increase as the number of rainy days increases, but also that this is not a linear (i.e. a straight line) effect, but one that accelerates as the number of wet days increases. This implies a higher *average daily rainfall* is associated with a higher *total number* of wet days per annum. This could be deduced from Table 2.3, but not so quickly, as the trend is partly hidden by records being in alphabetic order of station names.

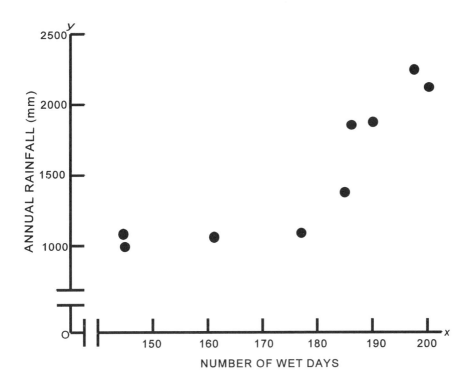

Figure 2.1 Graph showing annual rainfall plotted against number of wet days in 1978 for nine meteorological stations in Scotland.

Rearranging the data in increasing *x*-order would make the trend clearer, but still not as clear as in the graph. Exact values of *x*, *y* for each station cannot be read from this graph because of scale limitations, but approximate values (correct to about 2 significant figures) are obtainable.

If names of stations were written beside each point, the graph would become cluttered. If that information is needed, clutter may be avoided by putting a number beside each point with a 'key' to identify stations either on the graph itself, in the legend to the figure, or in the text. This is illustrated in Figure 2.2, where the key is on the figure. Some publishers prefer the key outside the figure. In this example I have used a convenient blank space on the graph for the key without making it look cluttered.

Whether Figure 2.1 or 2.2 would be used depends upon why the graph is needed. If you only want to show the nonlinear increasing trend in rainfall

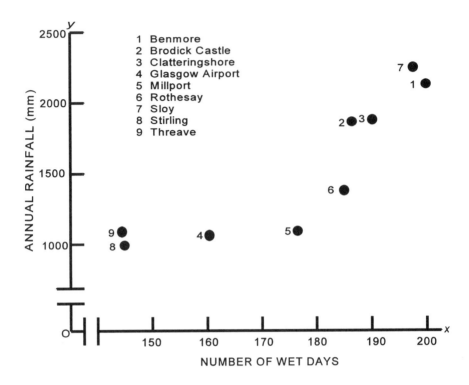

Figure 2.2 Graph showing annual rainfall plotted against number of wet days in 1978 for nine meteorological stations in Scotland. The number beside each point refers to the key at the top giving station names.

as numbers of wet days increase, station names are not relevant, and Figure 2.1 suffices. If names of stations are relevant, Figure 2.2 is better; these are (indirectly) relevant, for example, if one wants to see whether there is, say, a south to north trend in annual rainfall or numbers of wet days.

Edward R. Tufte gives a lucid account of the many uses of graphs in *The Visual Display of Quantitative Information* (FRIII).

Diagrams, other than graphs, and photographs have several uses:

- they often clarify descriptions of apparatus;
- they may help to explain the structure of organisms;
- they simplify *how-to-do-it* instructions.

Diagrams, including graphs, are often computer generated. Originals may be scaled down in printing, so, before preparing any, find out from your publisher what sizes are required; also whether labels or other text should be included, and if so, in what style or fonts. Pen and ink diagrams are best prepared by a professional draughtsperson; again seek advice from your publisher about format.

If you want to use photographs consult your publisher as soon as you have a contract, or even earlier. Colour illustrations are attractive but costly to reproduce, especially if good reproduction is needed to show detail or to ensure true colours. This could greatly increase the price of your book and threaten its commercial viability. A publisher may even want to limit the number of black and white photographs, for these may need to be printed on art paper if good reproduction is essential. Again, seek early advice, so that material is presented in appropriate form. Clear sharp transparencies are usually preferred for colour; large sharp prints for black and white, but technologies are changing rapidly.

2.2.5 References

You may find it strange that I suggest you consider this topic before you even start writing, feeling that a list of references is something you only need worry about when the main text is written. Experience has taught me that it is best to compile a list, and continually update it as you write, so think early about the form it will take. References to relevant published work are essential for most books about science or technology. Exceptions include some school texts, certain manuals, or popular books for nonspecialists. Details can be listed in three ways:

- as a **further reading** list;
- in a **bibliography**;
- as **references**.

Further reading lists may be placed at the end of each chapter, or at the end of a book. These are often annotated, and titles may be arranged in groups, based on the subject matter covered by each work. Not all books, papers or articles in these lists are necessarily referred to in the main text. Items in the lists provide background or supplementary information.

A bibliography lists all relevant reference material, whether or not it is cited in the text. Items are usually grouped in alphabetical order by author in a layout similar to that described below for references, although sometimes separate sections are devoted to books and to papers and articles.

References consist of a list of all work cited in the text as sources of information or evidence relevant to statements or assumptions made by an author. Sometimes, but not always, certain items from a further reading list may be repeated in the references. If you have both a further reading list and references, then a bibliography, which is rather like a combination of the two, is not needed.

There are two main methods for giving references. I prefer to give a reference by author(s) and year of publication in the text. This is called the Harvard system; further details are then given in the *references* (or the *bibliography*) at the end of the book, or sometimes at the end of each chapter. Chapter lists make it hard to find details if at a later date one only remembers there is a reference to Bloggs (1991) somewhere in a book, but cannot recall the chapter. I see no compensating advantage in 'end of chapter' listings, except when each chapter is written by a different author. Even while reading a chapter, it is as easy to turn to the end of the book as to hunt for the chapter-end to find details. Some publishers prefer an impersonal consecutive number system for references, known as the Vancouver system, in the form

It has been shown[2,3] that . . . or *It has been shown [2, 3] that . . ..*

Bibliographical details are then listed in numerical order at the end of the chapter or book. Resist this system if you can, for it makes it hard to find a paper by Bloggs unless you know the reference number. Sometimes references are numbered in alphabetic order of first author within each chapter, with full details in alphabetical order using this numerical key at the end of the chapter; this makes it easier to trace a reference, but it irritates authors, for if an additional reference to Bloggs is inserted at a late stage of drafting, many references will need to be renumbered and it even takes time to establish the correct insertion number. The main advantage claimed for numerical referencing is that it is less distracting while reading than a name and date system, but in practice I find this is more than offset by reduced utility for the reader who later wants to trace a particular reference.

With the Harvard system references in the text take a form

Smith and Jones (1985) showed that this process is reversible.

or alternatively,

This process is reversible (Smith and Jones, 1985).

If a reference is to a book it may be appropriate to indicate pages or chapter, e.g. (Smith, 1985, Ch. 3) or (Smith, 1985, pp. 17–22). Alternatively relevant pages or chapter may be included in the references section in the way indicated below in the examples for *Chapters in books*.

There are several ways of presenting details in a reference list or bibliography. Modern word-processing makes it easy to compile the list item by item as you make each citation. Follow a publisher's house-style (if one exists); alternatively, agree a style with your commissioning editor before compiling your list. Here are widely used ways of giving references to papers in journals, to books, or to chapters in books:

Papers
Woese, C.R. (1987) Bacterial evolution. *Microbiol. Rev.*, **51**, 221–71.
Vincent, W.F., Downes, M.T. and Vincent, C.L. (1981) Nitrous oxide cycling
.   in Lake Vanda, Antarctica. *Nature*, **292**, 618–20.

Books
Bergersen, F.J. (1982) *Root Nodules of Legumes: Structure and Functions*.
        Research Studies Press, Chichester.

Chapters in books
Sprent, P. (1988) *Taking Risks — The Science of Uncertainty*, Chapter 3.
        Penguin Books, London.
Reid, D.D. (1972) Does inheritance matter in disease? The use of twin studies
        in medical research; in *Statistics: A Guide to the Unknown* (ed.
        J. Tanur), pp. 77–83. Holden-Day, San Francisco.

These examples are indicative and do not cover all points in referencing; modifications to the format are legion. The date is not always put in parentheses; sometimes it is given after the volume number for journals, and the volume number is not always printed in bold. Names of journals may be abbreviated, or given in full. For papers, the paper title is sometimes omitted. If you use abbreviations, be consistent. Do not abbreviate the *Journal of the American Statistical Association* to *J. Amer. Statist. Assoc.* in one reference and to *J. Am. Statist. Ass.* in another. The former is a widely used abbreviation for that particular journal, although the latter uses the generally recommended abbreviation for *American*, i.e. *Am.*, and is therefore preferable. Some journals indicate at the head of each paper, or in a footnote, an approved citation abbreviation of their name.

Name all authors of a multi-author paper in the references or bibliography when there are four or less; when there are 5 or more sometimes all are named, but many publishers prefer the first three only to be named, followed by *et al*. In the text it is common to refer, say, to Smith *et al*. (1954)

if there are more than three authors; when there are two or three authors the usual practice is to write, e.g. Smith, Brown and Gray (1957), at the first mention and Smith *et al.*(1957) if that work is cited again, providing this will not cause confusion with another paper by Smith and others also published in 1957. Refer to two papers written by the same Jones in 1989 as Jones (1989a) and Jones (1989b). Use Jones, J.W. (1992) and Jones, R.M (1992) to distinguish between works by two different Jones published in the same year. For a reference to a dictionary or other work with no named author a form often used is Anon (1985), but cite under a named editor if there is one.

Although it is not often done, there is sometimes a case for including in parentheses at the conclusion of each entry in the references the numbers of the page or pages in your book where that reference is cited. I have found this particularly useful in books that describe a number of different procedures or techniques, for sometimes one remembers which author has described a relevant method but is uncertain under what subject heading it is presented in the text. If an author index, as distinct from a subject index, is included it is appropriate to include text page references for all cited authors in that index rather than in the references.

In this book I have given a fairly extensive list for further reading and only a few references. Quoted works are included in only one or the other. My criteria for listing a work under *further reading* are that I regard it as a useful source of additional information on certain topics, or as a work that many, or even all, authors should consult. Remember that not all items in a further reading list need be cited in the text and my list includes items not referred to elsewhere in the book. On the other hand, items listed in *references* are generally sources of quoted material or are works relevant to arguments I use or examples I give. A few sources of material, such as that in Table 2.3, are given fully in a relevant footnote so they need not be repeated in the references, though some authors like to do so.

It is tedious to get reference details correct and complete, but worth a lot of effort. You will frustrate readers if citations are incorrect, or if details of text citations are missing from both references or a further reading list. Be sensible about modifying the way you give references in text when this will help the reader. For example, I give in section 3.2.3 the names of some important reference books; this helps when the author or editor is 'Anon' or if the books are widely known by their title (e.g. *Who's Who, The Concise Oxford Dictionary)*. The reference I have just made to *Who's Who* is the only one in this book, and since I mention it only as an example of a well-known book and neither recommend it for further reading, or (heaven forbid) quote from it, there is no need for me to include it in the further reading list or in references. In later chapters I stress the wisdom of being consistent when there are choices; my advice here to modify sometimes the way you give references contradicts that recommendation, but I rate it a worthy exception, and a good illustration of the point that one should not be too dogmatic about rules when there is a good reason to modify them — a point I also make in the next chapter about some aspects of microstyle.

I consider other aspects of macrostyle (e.g. preparation of an index) in later chapters only because they are not relevant to the pre- and early-writing phases. An index is often part-prepared as you write, but in most cases it is only finished after a book is typeset; for practical reasons, usually at the proof-reading stage, or at the final submission of CRC.

3

Microstyle

3.1 GETTING THE STRUCTURE RIGHT

It is easier to write when you are confident about choosing proper words and putting them in the right places. You should be confident if you have been taught the skills of writing, or have read carefully some good books about style like those in the *Further reading* list starting on p.204. These explain, with differing emphasis and in varying detail, the good and the bad in microstyle — the role of grammar and good punctuation in bringing clarity, the importance of correct spelling, the lurking danger of verbosity.

If you are not sure whether you can cope with these matters, try this short self-assessment exercise. Read each of the ten numbered sentences below. Class as *unsatisfactory* any that you think are ambiguous, unclear, too verbose, or worded in a way that cries out for amendment by a competent copy-editor. Class those you think acceptable, even if you may prefer to word them slightly differently, as *satisfactory*.

1. The data, which were collected in 1991, show a marked improvement over that for the previous year.
2. When overheated, the operator should reduce the temperature by opening the valve.
3. Rat A exhibited an excess of weight over rat B.
4. The experiment was performed with two hundred day old rats.
5. We must face up to the need to try out new methods.
6. At this moment in time the laboratory is able to meet all demands and they expect to continue to do so in the future.
7. The male bird mated with 40 odd females.
8. Much of the recent literature is concerned with the summarization of the very many results that have previously been scattered widely throughout a large number of different publications.

9. There is considerable scope for gene manipulation of useful commercial attributes or for adaptation to new environments using *in-vitro* techniques.
10. Its input rate is fixed, though it changes periodically, under the two conditions of usage, at regular intervals.

If you thought most, or all, of these sentences were acceptable, I urge you to read a book, or books, about style before you start writing, and be ready to refer to guides on the proper use of words as you write. Fowler's *Modern English Usage* (FRII) or Gowers' *The Complete Plain Words* (FRII) are the kind of book all serious writers consult when they have a doubt about how to express something.

If you rated some of the sentences satisfactory, others unsatisfactory, you are a not very critical reader. Uncritical readers tend to be uncritical writers, so again I urge you to read more about style. Many good books tell you how to avoid stylistic nightmares like some in these examples, so I only discuss informally the grammatical boobs and misuses of punctuation that bedevil the works of inexperienced, and even sometimes experienced, writers.

Did you class all ten sentences as unsatisfactory and see sound reasons for doing so? If you did, you have a good 'feeling' for style. In the following comments the numbers correspond to those in the examples.

1. *That* should be *those*. *Data* is an awkward word. Strictly, it is the plural of *datum*; this is indicated by the plural verb forms *were collected* and *show*. However, there is a growing tendency to regard *data* as shorthand for *a set of datum*, and with this understanding, to use it with a singular verb associated with the word *set*. That convention would allow *The data, which was collected in 1991, shows a marked improvement over that for the previous year. That* is now correct. Some pedants refuse to accept this 'singular' use of *data*, but the history of language records many examples of changes in what is acceptable. The unacceptable aspect in this example is that the sentence treats *data* both as singular and plural. Some authors, and readers, may not like the use of the past tense *were collected* and the present tense *show* in the same sentence. This mixture is justifiable because collection took place in the past, but *show* refers to a characteristic which is as true today as it was when the data were collected.

2. It is implicit that the instruction refers to a valve on some equipment; but the sentence suggests the valve is to be opened when the operator is overheated! You may argue that common sense indicates that *overheated* refers to the equipment, not the operator, but the construction is similar to that in:

> *When full of water the operator may reduce*
> *pressure by means of the cock.*

Not only is there a *double entendre*, but the guidance is too vague to be useful. Just what is the operator to do? This version avoids ambiguity and is more informative:

> *When the tank is full of water, the operator may*
> *reduce the resulting pressure by opening the cock.*

If it is essential to take such action, write:

> *When the tank is full of water, the operator must*
> *open the cock to reduce pressure.*

It is important to be accurate and clear in technical instructions. If there is a possibility of an explosion if the operator does not open the valve, or cock, that should be made clear, or the supplier of the instructions may be legally liable for the consequences. People who write instructions often use weak words like *should* when *must* (perhaps emphasized as **MUST**) is what is needed. In this example even my amended version may leave you in trouble; you will need to alter *cock* to *valve* if you are to comply with the fashion for political correctness. I have more to say about this particular nonsense in section 3.2.7.

3. This is verbose. Why not *Rat A was heavier than rat B*? Verbosity in scientific and technical writing is common; it seems to be more evident to readers than it is to writers. This is a trivial example, but long-windedness is pervasive. Most people write too much at the drafting stage. Heed the message from Blaise Pascal:

> *I have made this letter longer than usual, only*
> *because I have not had time to make it shorter.*

4. This has three possible meanings. The experiment involved (a) 200 rats, each one day old; or (b) two rats, each 100 days old; or (c) an unspecified number of rats, each 200 days old. Ambiguity like this often arises when writers know what they mean without giving the matter a second thought. Maybe this writer's laboratory always experiments with day-old rats, but readers may not, often will not, know this. You might rephrase to avoid the ambiguity or else introduce hyphens, writing two hundred day-old rats (or 200 day-old rats), two hundred-day-old rats (or 2 hundred-day-old rats) or two-hundred-day-old rats, depending on the required meaning. A hyphen is part of the punctuation system; good punctuation aids clarity (section 3.2.2).

5. The fault is gratuitous appending of prepositions to verbs. Why not *We must face the need to try new methods*? The appendages are only minor irritants, but they are symptoms of verbosity. If you graded this example *satisfactory* do not be too worried, but the point is worth noting.

6. A bad attack of verbosity. Why not *The laboratory is able to meet all demands and expects to continue to do so*? Also, *laboratory* is (a collective) singular, so in the original *they expect* should be *it expects*. Avoid *at this moment in time* (or *in this day and age*); each is better replaced by *at present* or *now*. However, in this example neither *now*, nor the matching *in the future*, is needed.

7. This is ambiguous; *odd* could mean either *approximately* or *peculiar*. Use a hyphen, *40-odd*, or rephrase to *approximately 40* for the first meaning; change to *40 peculiar*, or *40 abnormal*, for the second.

8. Verbosity again. Specific faults are over-qualification of an adjective (*very many* when *many* suffices) and using a long phrase, *throughout a large number of*, when *in many* is adequate. I prefer

> *Much of the recent literature summarizes results*
> *scattered over many publications.*

9. *Using* in-vitro *techniques* refers to *gene manipulation*, not to *new environments*. Clarity is lost, or ambiguity may be introduced, by separating a qualifying clause from the statement it qualifies.

10. I had to read most of the paper in which this sentence occurred to discover that it meant:

> *The input rate alternates at regular intervals*
> *between two different fixed values.*

3.2 SOME CHARACTERISTICS OF ALL GOOD WRITING

Scientific and technical writing should obey rules that hold for all well-written nonfiction. Sound grammar helps clarity. The important rules of grammar are to do with ordering words and phrases and use of punctuation to avoid ambiguity and to make reading easier. Readers want to assimilate information without uncomfortable pauses induced by momentary doubt about what is meant or, even worse, ambiguity.

3.2.1 Catering for the fastidious

Some people do not take grammar very seriously because they are annoyed by a few pernickety rules that do little to help clarity. Such rules are only a small part of grammar. Nevertheless, it is wise to observe them (unless circumstances are exceptional), if only because failure to do so will annoy some readers. Minor crimes I have in mind include splitting infinitives and, less seriously still, ending a sentence with a preposition. Try not to commit these offences, even though it is widely accepted that to occasionally split an infinitive is permissible, and that a preposition at the end of a sentence is something most readers will put up with. That was blatant, I should have written: *It is permissible, occasionally, to split an infinitive, or to end a sentence with a preposition.* Do not, however, dodge the preposition problem with comic contortions like *A preposition is not a good word up with which to end a sentence.*

There are borderline situations where modern usage allows more than one construction, as with the word *data* discussed in section 3.1, example 1. A few readers may be irritated by:

Three were located in Kent, but none were found in Essex.

They will argue that *none* is a contraction of *not one*, making it the negative of a singular; so the correct construction is *none was*. But what about *almost none*?

Almost none of the original fifty were intact

is appropriate; *almost none* is now, in essence, a synonym for *only a few*; see, e.g. the entry for *none* in Fowler, in Gowers or in Jenkins (all FRII). Both Fowler and Gowers agree that *none* is now used more or less indifferently with a singular or plural verb form.

With words like *data* and *none*, where one construction may annoy pedants even though a case can be made for its use, I prefer the safer option, but your choice may be governed by your own usage, or your publisher's guide to style. Once you decide, be consistent.

Most sentences need verbs, but some in this book have none (see, e.g. the first sentences of my comments on examples 6 and 8 in section 3.1, where not using a verb adds emphasis).

Minor problems with idiom or convention may occur. *Your solution is different than mine* will jar on British, but not on American, ears. *Different from* is a preferred construction although *different to* is widely used, but not universally acceptable.

Sloppy expression is a more serious fault that often goes with grammatically incorrect construction. Example 2 in section 3.1 illustrated the problem in a simple context.

3.2.2 Punctuation

Publishers and editors complain that many — some claim most — authors cannot spell, and do not punctuate correctly. Punctuation is a cornerstone of style. In his *Nuts and Bolts of Writing* (FRI), Michael Legat devotes his opening chapter to punctuation; read his lucid treatment if you find punctuation a problem.

Most people cope reasonably with full points and interrogation marks; problems are more frequent with commas, semicolons, colons, dashes, hyphens, quotation marks and apostrophes; too little punctuation is more common than too much. Most of us find that, very often, we have to read a long sentence twice because we have, at first reading, misinterpreted the structure; largely because there is no punctuation to guide us. In the last sentence I used commas freely. Had I omitted the first comma and written *Most of us find that very often . . .* a reader might pause mentally at that point to ask *Find what very often?* Only after reading on would the intended meaning become clear.

It is a slight over-simplification to assert that there should be punctuation at any point where one pauses instinctively in speech when reading a passage aloud in a manner that conveys the intended meaning. Nevertheless, it is a good idea to read aloud anything you have written, and where you pause naturally, to consider whether punctuation is needed. What that punctuation should be depends, in part, on the nature of the pause. A short pause usually indicates a comma, a rather longer one a semicolon, a colon, or even a full stop. Some publishers disagree with what I have just recommended. They argue that punctuation should only be used if it is essential, claiming that it interrupts the reading process. I know what they mean, but I can assimilate what I read more easily if there is an indication of where a break, similar to an instinctive pause in speech, is appropriate.

Two examples in section 3.1 showed how hyphens may resolve ambiguity. This use corresponds to a shortening of the pause between spoken words. In speech this pause-difference distinguishes between *a blue-speckled bird* and *a blue speckled bird*. The former is clearly a bird of indeterminate colour having blue speckles, whilst the latter is ambiguous; if a blue bird with speckles of indeterminate colour is intended, it is better to write *a speckled blue bird*, perhaps even *a speckled blue-bird*. Similarly, a *large-winged bird* is clearly a bird with large wings, whereas a *large winged bird* might be either that or a big bird with a wing damaged by gun-shot. In words that were once hyphenated, the trend is to omit the hyphen if it serves no useful purpose and to use a single word, e.g. today, tomorrow, nonlinear, textbook (but you need a hyphen in un-ionized if it refers to electrochemical, not trade union, status).

Difficulties with punctuation may be accentuated if publishers have 'house conventions' that conflict with your instincts. A concise and useful guide to punctuation is included in *Hart's Rules for Compositors and Readers at the University Press Oxford* (FRII); known to the publishing world as *Hart's Rules*. The book is also a mine of information on matters like using quotation marks; when to use italics; how to hyphenate when splitting words at the end of a line; use of capitals; how to cite references; when to use the endings *-ise, -ize*. It also covers some conventions in scientific work, including biological nomenclature and setting mathematical formulae. Many publishers supply authors with guides to house-style on some, if not all, these matters. The *Guide for Authors* prepared by Basil Blackwell Publishers (FRI) covers practical points relevant to authors submitting to *any* publisher (including advice on how to approach publishers). If your publisher produces a style guide (Chapman & Hall, who publish the Spon imprint, does) that should take precedence over Hart's Rules; otherwise I recommend you follow Hart, although some of its critics complain of a bias towards Oxford University Press conventions.

An amusing punctuation quirk is the so-called *Oxford comma*; this is a comma before *and* in a sentence like

> *Mr Jones teaches mathematics, physics, chemistry, and biology.*

The comma is often omitted in the UK, but usually not in the USA, where it is better known as a *serial comma*. However, a comma is usual before an *and*, or before a *but* or an *or*, if what follows that conjunction has a different status to the earlier items in the list, e.g.:

> *Mr Jones teaches mathematics, physics and chemistry, but other science subjects only if required.*

Here the items listed before the word *but* are all single subjects taught by Mr Jones (so no Oxford comma before the *and*); what follows *but* is a group of subjects, not specifically named, that might be, but not necessarily will be, taught by him. Be careful about placing commas in sentences when some items in a series are paired and some are not. If you claim your favourite drinks are *whisky and soda, gin and tonic, beer and blackcurrant juice,*

I may consider your taste eccentric unless you confirm you mean

whisky and soda, gin and tonic, beer, and blackcurrant juice.

In scientific and technical writing punctuation in abbreviations, usually the full point, is important and I consider this in section 5.3.1. Closely related to punctuation is the way to present numbers. Should one write 10,000 or 10000 or 10 000 or spell it out as ten thousand, or use the hybrid 10 thousand? I discuss presenting numbers in section 5.3.2.

3.2.3 Spelling and choice of words

You may know the word you want, yet not be sure how to spell it. Spelling checkers in word-processing packages correct many errors, though no checker will bat its electronic eyelid at

Were where ewe too knight?

A checker does not tell you whether a word means what you think it does; there may also be inconsistencies in responses to English and American spellings. Standard checkers include few medical, scientific or technical terms, but supplementary lists are now available for these, or you may add words of your choice to most checkers. This is useful for technical or jargon words that you often use.

You need a good dictionary if you are unsure about the meaning of a word. For everyday words most well-known standard one-volume dictionaries are adequate. Several are available on disc (or do you prefer to write *disk*?), and may even be incorporated into word-processing programs. Increasingly, these and other reference books such as encyclopedias are also being distributed on CD-ROM. At a more sophisticated level *The New Shorter Oxford English Dictionary* (2 volumes) (FRII) and *Webster's Third International Dictionary* (3 volumes) (FRII) are useful; most writers have at least library access to one or both of these. *Webster* includes many scientific and technical terms and usually indicates both preferred English and American spellings. The *New* (1993) version of the *Shorter Oxford* contains more scientific terms than earlier editions. Subject dictionaries are available for most sciences and technologies; well-known publishers of these include *Oxford University Press, Penguin* and *HarperCollins* as well as many UK and USA general academic publishers. There are other books to help ensure you use words correctly. A companion to Hart's Rules is *The Oxford Dictionary for Writers and Editors* (FRII), and the more specific *The Oxford Dictionary for Scientific Writers and Editors* (FRII). The former will guide you on the use and meaning of foreign words and phrases, help with difficult spelling, with abbreviations and about use of capitals, etc.; the latter gives guidance for scientific terms; both are available on disc. At a more general level *The Economist Pocket Style Book* (FRII) and *The Times Guide to English Style and Usage* (FRII) are other useful references, as is the *US Government Printing Office Style Manual* (FRII). Many large companies and institutions publish their own manuals that give advice on spelling, punctuation and layout. A typical example is the *University of Minnesota Style Manual* (FRIII); as in most in-house manuals, some of the material in this is relevant only to that institution, but it also gives general guidance on spelling, punctuation and layout.

As it does for punctuation, Michael Legat's *Nuts and Bolts of Writing* (FRI) gives useful help on spelling, including a long list of words that are often misspelt; for some, Legat indicates differences between American and English spellings. When alternative spellings exist, be consistent. There are grey areas. I used to write *disk* for the thing I insert in my computer drive, and *disc* for the thing I put on the turntable of my record-player, but this was more habit than a meaningful distinction. In this book I follow Spon house-style, and a growing UK trend, and use *disc*.

All writers need a thesaurus, e.g. Roget (FRII), not only to help find a word with precisely the required meaning, but to avoid boredom from repeated use of the same word.

Several word-processing packages incorporate grammar checkers as well as spelling checkers and a thesaurus. Use grammar checkers with discretion, but they will often alert you to an unintended blunder. Mine keeps telling me, sometimes with justification, that my sentences are too long. It also produces a fair ration of nonsense, having particular difficulty in finding the subject for a verb when the two are separated by a qualifying clause. It wants to correct my *Seven sheets of paper were provided* to *Seven sheets of paper was provided*, insisting that *paper* rather than *sheets* is the subject. When I tried it on my *Were where ewe too knight* all it told me was that unless 'too' meant *also* or *excessively* I probably wanted 'two' or 'to'.

3.2.4 Sentences and paragraphs

Clear, short, grammatically correct sentences have a lot to commend them, but vary sentence length to avoid stilted or monotonous writing. Too many short sentences lead to *the-cat-sat-on-the-mat* sort of writing — a self-explanatory phrase used by someone who read a draft of this book to describe some of my text (now modified). Indeed, when presenting interrelated information, breaking one sentence into several may hinder understanding.

> *The vehicle has three wheels, a small one at the front and two larger ones at the back; the passengers sit directly over the back wheels, while the driver sits slightly behind the front wheel.*

is better than

> *The vehicle has three wheels. There is a small one at the front and two larger ones at the back. The passengers sit directly over the back wheels. The driver sits slightly behind the front wheel.*

What constitutes a paragraph depends upon what you want to say on a theme. When describing an experiment you might:

- list the components needed;
- explain how these are connected;
- how the experiment is performed.

Each aspect calls for at least one distinct paragraph. You may need a further split; for example, if some components are glass and some metal, each kind may be described in a separate paragraph. Several paragraphs may be needed to describe different stages of the experiment. Avoid excessively long paragraphs; few should exceed 200 words, the majority should be 100 words or less. Look at any long paragraph, and if it can be logically split into two, seriously consider doing so.

3.2.5 Holding readers' attention

To get, and hold, readers' attention present information in an orderly and carefully paced way; neither too rapidly nor too slowly. The former intimidates or frustrates readers; the latter bores or irritates. The ideal pace depends on the target readership. Readers are aggravated by:

- terminology or jargon that they do not understand;
- too rapid an introduction of new concepts, especially if these are not closely related;
- a pretentiously verbose style.

If you are conscious of such tendencies when you write, be careful not to over-correct; neither labour the obvious nor slow presentation with padding or by appearing to 'talk down' to readers.

If several new concepts are introduced, it may help if you illustrate each by a simple example. Judicious repetition, too, aids assimilation. If a difficult concept is introduced in Chapter 1, and not mentioned again until Chapter 4, readers may recall it only vaguely. Recap the basic idea in a sentence or two (more if needed). But be careful; you may bore, even insult, readers if you harp back to, or repeat, the trivial; at the other extreme, you will lose readers if you expect them to understand immediately, after a 200-word description, a concept that took you several weeks to master when you first met it.

Do not confuse relevant redundancy with verbosity. Redundancy is part of the communication process; at the simplest level it is built into grammar. In the sentence *Experiments confirm these results* the plural noun *experiments* and the plural verb *confirm* both indicate that more than one experiment has been performed. Here you see the perversity of grammar; to form plurals you add *s* to nouns and delete *s* from the present tense of verbs.

Timely reminders about, or repetition of definitions, or descriptions of concepts, are more sophisticated uses of redundancy. How much redundancy you need depends upon both the complexity of the subject and the sophistication of potential readers. Getting the balance right is not easy even when you are writing for a fairly homogeneous readership; it is hard for one with wide-ranging backgrounds and abilities. Writers of instruction manuals, self-teaching books, or of books on science or technical matters for the layperson should always bear this in mind. For the last of these categories, success may be more an art than a science. That it can be achieved in writing for a mixed readership is evident from the wide acclaim given Stephen Hawking's *A Brief History of Time* (FRIII) and the interest in Roger Penrose's *The Emperor's New Mind* (FRIII). Neither author talks down to readers; both books require careful reading even if you know something about the subject matter. The authors may have been surprised that they attracted so many readers. If you analyse their style you will see that both take great care in choosing proper words for proper places. Penrose is adept in his use of redundancy and often illustrates new concepts with an example, e.g., that for *Euclid's algorithm* on p. 41 of his book.

3.2.6 Consistency

Not only should your punctuation and spelling be correct, but, in these and other aspects of style, you must be consistent. Where there are acceptable alternatives, stick to one. Common trouble spots are hyphens, use of capitals, those *-ise* and *-ize* endings, formats for sub-headings and for references.

It is easy to write *fertiliser* in one paragraph, *fertilizer* in the next; *sand-dunes* on p. 7, *sand dunes* on p. 99; *latin square* in one chapter, *Latin Square* in another; *Smith and Jones (1979)* in one place, *Smith & Jones (1979)* somewhere else. Spelling checkers do not detect such inconsistencies. The differences seem trivial, but they become cumulative irritants for readers. If in doubt, follow recognized authorities like *Hart's rules, The Concise Oxford Dictionary,* Fowler's *Modern English Usage* or your publisher's house-style guide. Sometimes you feel you cannot win; I prefer *textbook* without a hyphen; so does the *Oxford Dictionary for Writers and Editors* and *Webster's Third New International Dictionary,* but prior to the 1993 edition, *The Shorter Oxford English Dictionary* went for *text-book.*

Problems also occur with use of *italic* or **bold**. Convention, or a publisher's house-style, may indicate a preferred form in many situations, but if this preference differs from yours, alertness is needed to achieve consistency. There are conventions about using *italic* in both botanic nomenclature (for genus and species) and in mathematical formulae and equations (for variables); if you are a botanist or a mathematician you will — or should — know the rules, but some are subtle (e.g. a^x, but e^x — the roman 'e' because it is the exponential constant, not a variable; I have been caught with that one). If you work in another discipline and need to refer to a plant or organism, or to quote a mathematical formula, check the conventions. Hart's Rules provides a basic guide. Do not over-use **bold** or *italic*. I use a lot of italic in this book, so I look to be ignoring my own advice. Apart from book titles, I use italic mainly to highlight quoted material or examples that illustrate specific points, and for referring to selected words from these. Because I give many examples, I decided italic was less intrusive than quotation marks, but you may not agree. Incidentally, quotation marks are not needed if quotes are displayed; the quotation should then be indented and put either into italic, or a different font or point size used. The terms *font* and *point size* are explained in Appendix C. Unless you are preparing CRC your publisher will make decisions about indentation, fonts and point size. If preparing CRC discuss such matters with your publisher before submitting your final copy.

3.2.7 Politically correct language

Legend claims that a *nice green lawn* was edited out of a children's story recently because some children have no such place to play. Political correctness cannot be ignored, even if that example, or a claim that a cockpit is no place for a nice lady pilot, make a nonsense of much of it.

In many countries, including some states in the USA, there are legal restrictions on the use of expressions that may imply that equal opportunities legislation is being ignored, or that may be regarded as incitement to civil unrest. Even in the less politically sensitive UK, the water authorities insist, much to the amusement of most plumbers and the public at large, that there shall be no references to *ballcocks* and *stopcocks*; instead, these must be described as *float-operated valves* and *stop-valves*. Also taboo are those old plumbers' stand-bys the *bastard file* and the *female joint*.

Publishers are very sensitive about the more understandable aspects of political correctness. In their *Guide for Authors* Chapman & Hall and Spon state:

> *It is our policy not to publish material which we deem to be sexist, racist or prejudicial in any way.*

Modern attitudes to gender and race demand care in the choice of words. *Gender* was once called *sex*, but that word, like *gay*, now has special connotations in general writing, though *sex* is still used with its old meaning by biologists.

A tacit understanding that, for pronouns like *he*, the male embraces the female, is no longer acceptable; escape routes are the cumbersome

> *If an author encounters this difficulty he or she must . . .*

or a twist to the rules of grammar (not recommended)

> *If an author encounters this difficulty they must . . .*

or restructuring

> *If authors encounter this difficulty they must . . ..*

The last is often the best way out. A way to avoid difficulties like

> *A nurse should monitor her patient's condition regularly,*

is to replace *her* by *the*.

Sensitivity about gender has spawned *chairperson, spokesperson,* etc. Use *humanity* or *the human race* rather than *man* or *mankind*. To avoid charges of chauvinism, or being sexist, when illustrating a point about, say, judges' placings, your publisher may prefer a skating or diving competition to a beauty contest. Be wary with adjectives like *rugged* and *pretty* that tend to be associated primarily with one gender.

> *You have to be rugged to climb mountains*

might suggest to some readers that climbing mountains is a predominantly male activity; so it is better to write, e.g.

You have to be fit to climb mountains.

Use scientific or acceptable popular designations for ethnic groups, or, where appropriate, an explanatory, non-derogatory generality like *many African races.* Prefer *third world* or *developing,* to *backward* or *under-developed* countries.

Some, but not all, such restrictions irritate me; but one irritated author is better than a dozen angry readers. More ludicrous manifestations of politically correct language like *child attendant* for *nanny* seldom impinge upon scientific writing, but as my comments about plumbing terms indicate, science and technology are not immune from the impact of political correctness.

3.3 ILLUSTRATIONS OF PROBLEMS WITH STYLE

Books from major publishers and research reports or papers in reputable scientific and technical journals contain only a few examples of glaringly incorrect grammar, or of poorly-constructed sentences. This reflects careful editing, for authors often deliver carelessly written manuscripts. In this section I indicate some faults in sentence structure. I consider problems with particular words and phrases in section 3.4. I have no wish to criticize individual authors, so I have reworded many real examples by changing the subject matter while retaining the stylistic fault.

3.3.1 Things that might have been put differently

Copy-editors may spare authors' blushes, but they cannot guarantee that all loose or ambiguous phrasing will be detected and corrected. An author may excuse, and a copy-editor pass, an ambiguous sentence like

The best estimators are complicated so that the solution cannot be obtained directly.

on the grounds that common sense will tell the reader it means

The solution cannot be obtained directly because the best estimators are complicated.

rather than

> *Somebody has, perhaps with malice, fiddled with the best estimators to complicate them in a way that makes it impossible to obtain the solution directly.*

Common sense may win, but given the author's original form, a reader will take a second or two longer to resolve the ambiguity. A more confusing example is

> *In order to better understand the complex structure and the nature of the reactions for this compound we decided to simplify the problem by breaking it down into smaller pieces which could then be studied individually.*

Was it the compound or the problem that the authors broke into smaller pieces? One must read more to find out. Loose or ambiguous statements slip through in any writing, but if you are habitually careless about expression, your writing becomes less acceptable to a reader (and to a publisher). Further, if you are careless about minor refinements you may find it harder to express difficult concepts as clearly as you should. It is often impossible, particularly when describing new concepts, to explain something without a passage that many readers will have to study more than once before they master it. Ambiguities, vagueness or verbosity make understanding harder.

For most examples in the rest of this chapter I give unsatisfactory versions in *italic* and preferred versions in **bold**. Some examples show poor constructions familiar to any widely-read scientist or technologist because, sadly, they do get into print. Others, selected from published work, but modified in the way explained above, are more specific.

I pointed out in section 3.1 that many books deal systematically with stylistic errors; this calls for excursions into the 'technology' of grammar. Kirkman (FRI) deals in this way with style for scientific and technical writing, while Palmer (FRI) discusses style in general writing. My approach is less detailed than, but more in the spirit of, Legat (FRI) who aims to give practical, rather than textbook advice, to 'someone who is trying to get work published'. Read Kirkman for his many examples and for explanations of the jargon of grammar; you will learn there of the pitfalls when you use *nouns as pre-modifiers* and sort out any confusion you may have between *pre-modifying nouns* and *the complements of prepositions*. I omit such formalities only because they are covered so fully by Kirkman and others. In many cases my suggested revision of a passage makes it self-evident where faults lie.

Remember that verbosity is often more evident to readers than it is to writers. Here are some examples with, usually obvious, simplifications:

Much of the relevant literature is concerned with the summarization of . . .

Much of the relevant literature summarizes . . .

. . .it may be the case that event times of individuals are associated with . . .

. . . event-times of individuals may be associated with . . .

Event-times is jargon; my inclination is to spell it with a hyphen; you may not agree.

It is required that an object be allocated to . . .

An object is allocated to . . . or **We allocate an object to . . .**

I comment in section 3.3.2 on over-use of the passive form.

It is precisely this procedure with which this chapter is concerned.
This chapter is about this procedure.

Record this code and relay the information to the service technician.
Note this code for the service technician.

If you require more space for storage than a single unit . . .
If you need more than one unit for storage . . .

The degree of complexity of the influencing factors covering phase changes during transmission is so complex that . . .
Factors influencing phase changes during transmission are so complex that . . .

The operating system was examined to ascertain whether it was achieving the original aims detailed at the start of the project.
The operating system was examined to see if it achieved the original aims.

The first version contains too much redundancy; also 'to ascertain whether' is better shortened to 'to see if'.

Over-use of abstract nouns formed from verbs often leads to verbosity. Instead of

Immobilization of the particles results from the addition of acid.

write

Adding acid immobilizes the particles.

The latter gets rid of the awkward prepositions *of* and *from* that creep in when using abstract nouns with verbal origins.

> *Cost savings were brought about by the introduction of new insulating material.*
> **New insulating material reduced costs**.

Here changing from passive to active and dropping the (unnecessary) verb equivalent *introducing* of the abstract noun *introduction* both help.

In section 3.1 I mentioned the need for accuracy and precision; it is bad style to be imprecise, especially when this amounts to inaccuracy. Technical writers are sometimes reluctant to be positive or specific. A change of words is often all that is needed.

> *When the danger light comes on the operator should reduce pressure by means of the regulating valve.*

is only an encouragement to do something vague. What is almost certainly meant is

> **When the danger light comes on the operator must reduce pressure by opening the regulating valve**.

3.3.2 Active or passive

Passive, rather than active, verb forms used to be common in scientific and technical writing. The practice, once encouraged in professional and learned journals, stemmed from an erroneous belief that the passive form was needed for objectivity. The passive is sometimes appropriate, but it often leads to stilted, or monotonous, prose. Most journals now recognize that writing may be objective without being impersonal. Here are two examples with preferred alternatives:

> *It is necessary to understand . . .*
> **We must understand . . .**

> *In this book the view is taken that . . .*
> **We take the view that . . .**

I have used *we* in the active form. *I* is often appropriate. An objection to *we* in scientific and technical writing is that it savours of the royal *We* and may give an impression that the author is talking down to the reader. This may be so if an author writes *We have established . . .,* when clearly meaning *I have established . . ..* However, using *we* is proper in papers or books by joint authors who are reporting their own work or expressing jointly held views. It is also acceptable if used to imply a partnership with the reader in sentences like *We would be unwise to accept this as the sole cause.* Whatever you decide, be consistent; don't jump from *I* to *we* and back again when only one or the other is appropriate.

3.3.3 Order

Reordering words or phrases may remove ambiguities and save readers seconds in grasping the intended meaning, because ideas are then arranged more logically. Here are examples:

> *. . . described in work as yet unpublished by Jones.*
> **. . . described in work by Jones (as yet unpublished).**

The original implies work (that might have been done by anybody) has not yet been published by Jones; the revision tells us, as intended, that Jones did the work, but has not yet published it.

> *. . . is the measurement relating to the outcomes in two ears of two subjects in a given family, conditional on the outcomes in their other two ears . . .*
> **. . . is the measurement for the outcomes in one ear of each of two members of a family, conditional on the outcomes in their other ears . . .**

This is no longer and avoids visions of 'four-eared' monsters.

> *Use a dual Y-axis plot.*
> **Use a chart with two Y-axes, one on the left, the other on the right.**

The first form uses non-standard jargon — dual *Y-axis* — the meaning of which is made clear in the revision. Writers of computer software manuals often, unjustifiably, expect readers to know 'in-house' or 'nonstandard' jargon.

> *At the data entry line enter the name of the file in which you wish to save the selected block or from which you wish to load and insert at the current cursor position.*

At the data entry line, enter either (i) the name of the file where you want to save the selected block, or (ii) the name of the file from which you wish to load a block to be inserted at the current cursor position.

The extra words and punctuation clarify the possible actions.

Through the planned introduction of new components into selected cylinders, insight into the temperature changes, and therefore into the stresses leading to metal fatigue has been forthcoming.
Planned introduction of new components into selected cylinders has given insight into temperature changes and the stresses leading to metal fatigue.

The original version is not misleading, but separating 'insight' from the qualifying 'has been forthcoming' is unwieldy; it takes more effort to understand the passage.

All studies to date suffer from the caveat that the effect of any resulting emission on the global structure of the protein, and thus any long range consequences, cannot be predicted.
We cannot predict from studies to date either the effect of any resulting emission on the global structure of the protein or any long term consequences.

In the first version *caveat* means *proviso*, which is acceptable, although the more common meaning is *warning*. If *proviso* is intended, it would be better to say *are subject to* rather than *suffer from*. The alternative version is more straightforward.

A major requirement when developing a market for a new drug is a clear and definitive statement of its potential, giving information on its therapeutic value and warnings about possible side effects.
When developing a market for a new drug, we need a clear and definitive statement of the drug's potential, giving information on therapeutic value and warnings about possible side effects.

Ambiguity arises from using *it* in a context where *it* might refer either to the *drug* or the *market*. In this example the subordinate clause *giving information on . . .* makes clear that *it* refers to the drug; but given the first version, if the reader has previously decided *it* refers to the *market*, a mental switch is needed. The revised form makes clear that the *definitive statement* refers to the *drug*.

3.3.4 Misinformation and trouble with numbers

Wrong use of a word can make a statement meaningless.

> *The equations*
> $$4x+3y=5 \text{ and } 40x+30y=50$$
> *are inconsistent.*
>
> **The equations**
> $$4x+3y=5 \text{ and } 40x+30y=50$$
> **are equivalent.**

This illustrates a common fault in instruction manuals where writers have difficulty with the jargon of a particular subject (here mathematics) which is not the core subject of the writing. The equations are equivalent, and therefore *consistent*, although there is not a unique solution. If the second equation had been $40x+30y=70$ it would have been *inconsistent* with the first.

Writers often run into trouble with quantitative ideas. Because of the speed with which a newspaper is produced, one might excuse *The Times*, a paper normally careful about such matters, for reporting on 11 August 1993 that

> *The Lord Chancellor has opened the way for a move towards the American style 'no win no fee' system of legal work by agreeing that lawyers may double their fees if they take on a case for nothing and win it.*

A lawyer should not be happy about doubling a fee of nothing! What is intended is a doubling of the fee that would normally be charged; insertion of a word such as *normal* or *usual* before *fees* in the penultimate line would clarify the position, though I would prefer to reword as

> **The Lord Chancellor has opened the way for a move towards the American style 'no win no fee' system of legal work by agreeing that lawyers may take a case on the basis of double the standard fee if they win, but no fee if they lose.**

3.3.5 The cumulative effect

Some examples in sections 3.3.1 to 3.3.4 are trivial; but careless writing, or failure to give due thought to style, often leads to a global lack of clarity that permeates a paragraph, then a section, next a chapter, and finally a whole book.

The extract below is from a book I wrote 25 years ago (Sprent, 1969, p. 30). My target readers were graduate or advanced undergraduate statisticians wanting to learn more about a special area of statistics. The quoted paragraph requires some statistical knowledge to grasp fully the point I was making, but I think even non-statisticians will see the need for improvement, as I did on re-reading it some quarter-century after I wrote it. I hope you agree that the version in **bold** is an improvement.

Nature herself has altered relationships between sizes of parts of organisms in the evolutionary process, and these changes can be studied. We might find a linear relationship, perhaps after taking logarithms, to hold fairly well between measurements of height above the ground of the top of dogs' heads and their body length, nose to tail, more or less irrespective of their breed, or lack of breed; but there would be notable exceptions such as dachshunds. Suppose however that we confine our attention entirely to dachshunds; we would probably find that within that breed a linear relationship held between the two measurements, apart of course from superimposed random variability representing individuality. This relationship would be different from that holding for more conventional breeds; it is also possible that we may find within other breeds linear relationships, sometimes similar to that holding in general between breeds (other than exceptional ones) but sometimes rather different. This example indicates an important point of principle — namely that some of the departure variability associated with a relationship may be due to differences in the relationship as we move from group to group (groups corresponding to distinct breeds of dogs). Differences in the relationships within groups may only be clearly observable when there is extreme departure from the between groups pattern, but the concept of two types of relationship — those holding within and those holding between groups, is important.

Nature may alter the relationship between sizes of parts of organisms; this is part of the evolutionary process. For example, there is a 'nearly linear' relationship between (a) the logarithms of measured heights above the ground of dogs' heads and (b) logarithms of their body lengths. If we plot such data for many dogs of different breeds (we call this the pooled data), we find considerable scatter about the 'trend' line of best fit. This scatter arises from (a) differences in the growth patterns of individual dogs *within* any one breed and (b) differences *between* the average

growth patterns for different breeds. For example, dachshunds have longer bodies, relative to their height, than do most other breeds. If we plot separate lines of best fit to our data for each breed, most, if not all, such lines will differ from that for the pooled data. Also, for any breed the scatter of individual plotted points about the trend line for that breed will usually be appreciably less than the scatter about the pooled data line. The differences between the 'pooled' and the individual-breed trend lines reflect variation *between* breeds arising as part of the evolutionary process. Departures shown by individual dogs from a trend line for all dogs of that breed reflect individuality *within* a breed.

The concept of two types of departures from a pooled relationship — those *between*, and those *within*, groups (here breeds) — is important.

In the revised version I have taken more care with grammar and punctuation. The argument is put in a more logical order. In the original, confusion results from jumping between the trend line for the pooled data, that for dachshunds, those for other breeds, and departures from these lines. Irrelevant remarks about dogs of mixed breeds are left out in the revision. Even this revised form, read in isolation, uses undefined terms like *lines of best fit*. These points had already been clarified in the book.

3.4 OTHER ASPECTS OF STYLE

3.4.1 The right word

Word order is important; but even more fundamental is the choice of each word. I drew attention in section 3.3.3 to possible ambiguity in the use of *it*. Avoid pronouns if they trigger ambiguity.

The screws should be removed from the electrodes. They should be stored for possible re-use.

is ambiguous.

The screws should be removed from the electrodes. The screws should be stored for possible re-use.

removes ambiguity, but to avoid the same start for consecutive sentences, prefer

The screws, after removal from the electrodes, should be stored for possible re-use.

Repetition of identical words or phrases is tedious. Also, it may cause reading difficulties, because the eye jumps easily to the wrong point if the same word appears at much the same place in consecutive lines, or at the beginning of nearby sentences. Using a thesaurus to find synonyms is one solution, rephrasing is another. Many word-processing packages have a built-in, but often limited, thesaurus.

Rather than

Satisfactory functioning depends on careful control of the power input. Controlling power input to within 1 per cent of the optimum is usually satisfactory. If performance is still not satisfactory adjust the control marked 'Stabilizer' until a satisfactory performance is achieved.

write

Satisfactory functioning depends on careful control of the power input and is usually obtained by holding this to within 1 per cent of the optimum. If performance is still inadequate, adjust the control marked 'Stabilizer' until an acceptable standard is achieved.

However, when talking about specific objects or procedures, do not confuse readers by giving them new names simply to avoid repetition. Write:

We carried out greenhouse experiments and field trials for the first three varieties, but only a field trial for the remaining variety.

rather than

We carried out greenhouse experiments and field trials for the first three varieties, but only an outside test for the other crop.

In the latter form it is not clear that *outside test* and *crop* are to be interpreted as equivalent to *field trail* and *variety*.

Avoid clichés — particularly platitudinous ones like *It is well-known that* . . . unless they are relevant. By all means say *It is well-known to mathematicians that there is no construction for trisecting an angle using only a straight edge and compasses*. Do not use the phrase as a substitute for *It has been proved somewhere, but I can't remember where*. Another over-used phrase is *To name but a few* . . . . It often suggests that the writer is sure there are more, but cannot think of them. The phrases *It has often been observed that* . . . , *It is our intention to show that* . . . are platitudes that generally have no useful information content, contributing only to verbosity.

Avoid a long word when a short one will do, e.g. *commence* for *begin*; do not use a long phrase when a word or shorter phrase suffices. Blake and Bly (FRI), pp. 80–1 give over 40 examples, and follow these by a list of phrases that are usually best omitted altogether. Follow their advice and write, e.g., *soon* rather than *in the near future* or *at your earliest convenience*. Avoid also genteel euphemisms. If half the rats were killed in your experiment, say so, rather than claiming that *half the rats were sacrificed*, as I recently did.

Most people have a subconscious tendency to use frequently one or two words or phrases that are not needed or could be replaced by something else. Reading aloud what you have written often exposes this. In the several redrafts of this book I have removed *Note that* from the start of more than 100 sentences and also scrapped the word *indeed* from several dozen others.

Be careful with idioms and colloquialisms. The wide use of English in science and technology means a number of your readers, even a majority, may not speak English as their 'first language'. Even in predominantly English-speaking countries — Britain, USA, Australia, much of Canada — words and idiom are used differently. Although British idiom and usage may jar in the USA, and USA idiom and usage may jar in the UK, scientists are generally prepared to accept transatlantic variations like *gotten* and *got*, *specialty* and *speciality*. However, some words, especially colloquialisms, may not survive an Atlantic crossing, and will almost certainly confuse readers for whom English is not the first language.

These problems are most acute when writing for a nonspecialist readership, but they turn up even in texts at the graduate or undergraduate level. I recently described a hypothesis set up primarily to test whether there was sufficient experimental evidence to reject it as a *cockshy*. A translator into French asked me to explain this, but we found no obvious French equivalent. If professional translators face such problems, pity the reader for whom English is not a first language? Bear these readers in mind as you write; also, when choosing examples, avoid contexts that may be unfamiliar to readers in another country. A British author of a textbook on dynamics who looks for good American sales, should keep cricket balls out of examples on the behaviour of projectiles. My attempt to explain cricket to an American on a transatlantic flight only bore fruit when I made the *bowler* a *pitcher*. Likewise, if you are an American writing about random variation, keep off baseball scores in your examples if you are looking for good European sales.

Beware of different usage of everyday words if you are writing technical instructions, operating manuals, etc. for use in several countries. Here the important person is the target reader. You may even need to produce separate sets of instructions for each side of the Atlantic. While *got* and *gotten* may not confuse, word equivalents like *faucet, tap; lift, elevator; gas, petrol;* may cause difficulty.

Watch also words with two meanings but the same spelling. These are manna only to compilers of crossword puzzles, letting them dream up clues like *Very small time (6)* for the solution *minute*. So, although *I like four minute eggs for breakfast* probably means I like eggs that have been boiled for four minutes, it could mean that I like four very small eggs. A hyphen, *four-minute,* would help here. The problem just described does not arise in speech because of differences in pronunciation, but, whether spoken or written, the instruction *Without raising the feet lift the bottom off the bed* (I did not make that one up!) leaves doubt whether you are to manoeuvre part of your anatomy away from the mattress or to start dismantling the bed. Watch out when writing about a specific subject where an everyday word is also used with a technical meaning. The sentence *The most significant aspect of this finding is that it is not (statistically) significant* is confusing and becomes meaningless if the word 'statistically' is left out.

3.4.2 Units

Not all scientists approve the choice of some of the base units, e.g. the Pascal, in the SI system of measurements, but the system is widely used for communication between scientists because it eases the problems of comparing data from a variety of sources, avoiding the need to look up conversion factors between, say, imperial and metric units, with the danger of applying those factors incorrectly!

However, in the United States, especially in technical reports and communications addressed to a broad readership, imperial units (or similarly named units such as the US gallon) are widely used. Internationally, too, one always refers to 3.5-in. and 5.25-in. computer discs.

Avoid also SI units in contexts — usually in technical writing related to applications — where non-SI units may be appropriate because of customary usage by your target readers. For example, in aviation an astounding mixture of units is used internationally. Air speed is measured in knots, altitude in feet, airport visibility in metres, distances in miles or nautical miles (or sometimes kilometres), atmospheric pressure in millibars (in inches of mercury in the USA!). It is perhaps surprising that there are so few mid-air collisions, but using SI units when writing for pilots or air-navigators might increase the risk.

If you feel some of your readership would like SI and some imperial or other units, give your information primarily in one set with conversions to the other in parentheses. When converting, avoid spurious accuracy from quoting too many decimal places, e.g. put '1.3 in. (33 mm)', not '1.3 in. (33.02 mm)'.

4

Seeking a book publisher

4.1 DRAFTING A PROPOSAL

4.1.1 The first approach

If you are planning a book and you have decided what kind it is to be and are confident about style and related matters, start writing. A sensible first step — certainly an early one — is to channel some of that writing towards a proposal to send to a publisher. This helps to clarify your thoughts, though that is not the main objective. If you followed the advice in section 1.8 you should have one or more publishers in mind. Until recently, simultaneous approaches to publishers were frowned upon. Some publishers now do not mind multiple submission of a proposal **providing** they are told you are doing that; but, at risk of being dubbed old-fashioned or too gentlemanly by half, for nonfiction I recommend offering a book to only one publisher at a time. If a publisher is lukewarm about your idea and knows you are submitting the proposal elsewhere, this may swing the decision against you.

There are no hard and fast rules about the best form of approach, but several 'don'ts' emerge in this chapter.

You might send a preliminary letter outlining the type of book you are planning, giving a provisional title, and stating why you think the book is needed (i.e. indicating the potential market) and how it might slot into the publisher's list. This last could be to fill an apparent gap in a series, or because the firm traditionally publishes books on your topic. Outline your qualifications for writing the book. Be brief and to the point. The publisher is likely to write back asking either for a synopsis and one or two draft chapters so that your proposal can be considered further, or to express regret (with varying degrees of sincerity) that the firm cannot consider your proposal, often giving one of these reasons:

- your proposal resembles too closely a book already on the firm's list;
- the firm has recently accepted a book on this subject from another author which overlaps your coverage;
- the firm does not think there is a place for your book in the present market;
- the work does not conform to the firm's current publishing plans;
- because of a change in policy, the firm is no longer publishing works of this type.

A preliminary letter that brings a quick negative response leaves you free to approach another publisher immediately. A positive response will almost certainly ask for a synopsis and for some specimen chapters. If you are fairly confident that a publisher you are approaching is likely to be interested you might skip the preliminary letter and send a synopsis, draft chapters and other information straight away. You lose on postage if the material is rejected (always include stamps to cover return postage with an unsolicited submission), but any negative response is likely to come almost as quickly as it would to a preliminary letter. If the draft material is good, an editor who might say 'no' on the basis of a preliminary letter may be willing at least to enter into a dialogue.

4.1.2 Be professional

Prepare your submission carefully. All publishers complain that much of what they receive lacks style, is peppered with spelling errors and is badly punctuated. Publishers like authors to be professional. Your final manuscript (although never hand-written, that is what your typescript, or word-processor printout, is called) will be very different from an early draft, but that is no excuse for a sloppy proposal. Typewritten work is acceptable, but most serious writers now use a word-processor. The print quality should be typewriter standard or better. Draft quality from a dot-matrix printer, especially if the ribbon is worn, irritates publisher's editors — and everyone else; near letter quality (NLQ) output is just acceptable. Use a daisy wheel (now rather old-fashioned), or a laser or inkjet printer if possible. If you do not own a good printer, you may have access to one where you work.

Both synopsis and specimen chapters (and eventually your final manuscript), should be typed on white A4 paper (210×297 mm) of at least 70 g/m² weight (bond), but 80 g/m² is even better. These are European specifications; equivalent standard USA paper size differs slightly. Keep copies (photocopies or carbons) of everything you submit. Text should be double spaced with at least 25 mm (1 in.) margins at the top and the bottom of each page and not less than 30 mm, or better 38mm (1.5 in.), margins at each side. Print on one side of the paper only. Modern word-processors and printers let you use several fonts (typefaces) at many point sizes. Don't get carried away by this facility. Choose a font, preferably either Courier or Times Roman (which are available for most printers), as standard. A good size is 12 point, but 10 point is acceptable with a high-quality printer. Providing you do not make too many corrections, these may be clearly handwritten at the appropriate points in the text, putting them between the lines rather than in the margins. Tables, graphs and diagrams are best collected separately, one to a sheet; tables in one bundle, graphs and diagrams in another. In the text the approximate position for each may be indicated by inserting

[Table 2.3 about here]

or

[Figure 7.2 about here]

as appropriate, at a suitable point in the manuscript. Alternatively, indicate the position by a written marginal annotation.

Most publishers prefer manuscripts to be left-justified only, i.e. the right margin is left ragged; this helps them assess the length of a book. If possible, follow this convention with the specimen material, but you may have technical reasons for using full justification (both right and left), and editors are unlikely to be put off by this in sample chapters, but they may ask for the final manuscript to be left-justified only. Avoid the temptation to use a range of point sizes or typefaces for headings, sub-headings etc. Only use bold or italic if these are really needed and if your printer reproduces them clearly. With dot-matrix printers, bold or italic are not always satisfactory and it is better to single-underline anything that is to be printed in italic and to put a wavy line under anything to be printed in bold (these are standard markings for typesetters). If using daisy wheels, you will need to change wheels for italic. Most publishers are happy if you:

- use one font and point size throughout;
- use double spacing for all material including headings;
- underline to indicate essential bold or italic;
- start each chapter on a new page.

If possible, keep the same number of lines on each page, except for the last page of a chapter. This makes it easier to estimate the length of a book. Word-processor word counts are not very good for this, especially when tables and formulae are involved. Number pages consecutively — do not start afresh at page 1 in each new chapter. Include a title page, giving the book title, your name (and those of any other authors if it is a joint-authorship book), address, telephone number, and, if relevant, fax and e-mail numbers.

If you would like to submit the final version as camera ready copy (CRC) say this when you submit your proposal. The editor will tell you whether that is acceptable, and may require evidence that you can do the job to meet house standards. Not all publishers accept CRC. Similarly, if you would like to submit your work for typesetting from disc, indicate this at the time of submission. Some publishers prefer typesetting from disc, some accept it grudgingly, others refuse to consider it. Nearly all who accept discs specify a strict format for these. These possibilities are discussed further in sections 5.2.2 and 5.2.3. There are indications that submission on disc will increase in popularity in future.

4.2 WHAT TO PUT IN YOUR PROPOSAL

Your proposal should:

- say what your book is about (outline or synopsis);
- give your qualifications for writing it (in a covering letter or in a *curriculum vitae*);
- include specimen chapters;
- give, if possible, an estimate of the proposed length, usually expressed in number of words. Depending upon page size and type font a 200-page book usually contains between 60 000 and 75 000 words;
- indicate the target readership and potential market you have in mind;
- list any strong competitors to your book, indicating why you think yours is better;
- give a realistic indication of when you expect to be able to deliver the completed manuscript.

You are offering goods in what has, in recent years, been a buyers' market. Do not make your synopsis too long. One, or at most two, paragraphs for each chapter is enough. For Chapter 2 of this book an appropriate synopsis would be:

Chapter 2: The theme is macrostyle. Points covered include: (1) the target readership and its influence upon style; (2) the importance of an attractive layout and ways to achieve this; (3) division into headed sections and subsections illustrated by examples; (4) how best to use tables, graphs, diagrams, illustrations, etc.; (5) ways of citing references.

Your qualifications are often best given in a covering letter. If this is much more than one A4 (single-spaced) page you are probably being verbose, or have an inflated idea of your importance. A (usually abridged) *curriculum vitae* (CV) may be appropriate if you have special qualifications for writing the book; if relevant, a list of your publications may help (the editor might want to look at some of these); outline your teaching experience if you are writing a textbook.

Send only two or three specimen chapters. These should include the introductory chapter, which, as Blaise Pascal pointed out (section 1.1), is often difficult to write. Indeed, it may pay you not to write this chapter until you have completed the other specimen chapters and the synopsis. It may then be easier to see the best starting point. I selected the Pascal quotation for possible use soon after I began to write this book, but I had drafted five chapters before deciding to start the book with it.

The introductory chapter is often used to indicate the general nature of a book. Do this concisely; resist the temptation to make it a résumé of the entire work. For many books a short introductory chapter is best, but sometimes (particularly in textbooks) the first chapter might jump straight into basics and cover one key aspect of a subject; if so, the word *introductory* is no longer appropriate.

When I submit a proposal I often send the first two chapters; if these do not fully reflect the flavour of the book I add a draft of a later chapter that I regard as representative. Draft material may seem disjointed if you have to give cross-references to material that is not yet written, but a publisher's editor will understand this.

The information about what you see as the market for your work is important; give this concisely, e.g. is it for undergraduate students and research workers, or the interested layperson (or perhaps all of these)? Is it for a specialist readership (e.g. research workers in information technology, design engineers, cell biologists)? Mention, too, if there are particular overseas markets where it may appeal. Remember that for UK authors the USA is an overseas market and *vice versa*. If you know there will be a potentially large market in Kenya, India, Japan or Australia, say so. Some of these things may seem obvious to you, but the publisher may not know them. Also, some publishers concentrate their marketing in specialized channels (e.g. direct sales to educational authorities and schools or to libraries) or in certain countries or territories. This implies that if you or the publisher do not see your book having potential in the relevant markets, time may be saved by both parties if this becomes obvious at an early stage.

Listing the competition helps the publisher assess the viability of your project. It also shows that you have studied the market and this indicates your professionalism. Include in your list not only books already published, but if you know that any other relevant books are being written you should say so. Explain clearly in what way you think your book will be superior to competitors. Do not over-state your case, but if the other books you list have shortcomings explain how you intend to overcome these in your work and also highlight anything that is novel in your approach.

Be realistic when giving a projected completion date for your manuscript. It will probably take longer than you think. Full-time writers often spend 12 months or more on one book. The exact time will depend on the subject matter, how much background research is needed, how much polishing of style is appropriate. If you are a part-time writer think carefully about how many hours you will be able to devote to your writing each week; or if you are, say, an academic, and only expect to be able to write seriously during vacations, assess your timescale accordingly. As a rule-of-thumb I suggest you try to make a reasonable estimate of how long it will take you to write the book and then **double** that time to provide your stated assessment of when you could submit your manuscript. I am being realistic, not cynical. If you do not intend to start serious writing until you have a contract, indicate in your proposal that your estimated time to submit the manuscript is from the date of signing a contract.

Unless a publisher rejects your proposal almost immediately as unsuitable, expect to wait four to six weeks for a response. If you have heard nothing after two months, send a polite note asking when you might expect a reply. If nothing is heard within a further month, ask for the return of your material. A delay in response from a publisher sometimes mean the firm is seeking an 'outside' opinion on your proposal, but firms do vary in efficiency; you need some patience because a major publisher is always considering a large number (sometimes hundreds) of proposals.

If a publisher shows interest in your work there is often further preliminary discussion or correspondence to clarify various points. A publisher's editor who doubts the commercial viability of your project may suggest ways of improving this; or might advise you to submit your material to some other named publisher. If your proposal does not find favour you will probably be told why; the reason will often be one of those listed in section 4.1.1. If this happens try another publisher.

Consider carefully all suggestions for improving commercial viability. Keep talking if you feel you could meet most or all of these, and still produce a book which you are happy to write. If suggested alterations hold no appeal, you may be wise to submit your proposal elsewhere. Publishers sometimes make inappropriate suggestions, but if several publishers suggest similar amendments, the odds are that all are right.

If a publisher expresses serious interest ask how, if the firm publishes your book, it will be marketed. Good publishers are usually happy to tell you, because they are proud of their marketing arrangements. It may well emerge from discussions that although the publisher has good arrangements in the British and European markets, the firm has little interest in the American market. If you see the latter as the main market for your book you will probably decide at this stage that you need to look for another publisher or else ensure that contractual arrangements allow you to sell American rights separately. In general, my advice to a UK author in these circumstances, would be to seek another publisher. A colleague who is closely in touch with the American scene assures me that US writers are more 'on-the-ball' than their British counterparts when it comes to checking whether a potential publisher's marketing arrangements are appropriate for a projected book.

4.3 PUBLISHERS' CONTRACTS

4.3.1 The draft contract

If all goes well you will be offered a contract. Publishers' contracts cause more misunderstanding and anguish — some of it long term — than most facets of authorship. Even before you get a draft contract, a publisher's commissioning editor may outline the main terms, specifying when the manuscript is required, a royalty rate, and whether an advance payment will be offered.

The terms may sound attractive, but if you are a first-time author, uninformed about the workings of the book business, keen to see your work in print, beware of potential pitfalls. There is generally no need to get bloody minded at this stage, but do not rush in. A preliminary offer does not always mean what you think it does. I am not implying that publishers regularly try to pull a fast one. Publishing is big business with a high risk element; only rarely does a publisher make a large profit from any one book, especially one by a first-time author on a specialist topic. Indeed, the publisher may expect to subsidize your book with profits from one of a few money-spinners on the firm's list. This may be a factor in shaping the offer to you. A commissioning editor costs each project carefully before making an offer. When making one the editor may assume (often wrongly) that you understand its meaning, or that you will take advice as to its suitability.

You will be told the offer is subject to contract, but the main points may be passed to you in terms like:

We would want the completed manuscript in nine months' time; there would be a £300 advance against royalties based on the usual paperback rate of 7.5% on home sales and 10% on export.

Potential traps lie in whether the delivery time is realistic, what is meant by *usual,* and failure to state the base for royalty rates. If you said in your proposal that it would take you 18 months to write the book, then nine months is unreasonable. There is no such thing as a *usual* royalty rate common to all publishers; many pay a royalty of 7.5% *of the retail price* for paperback sales on the home market; but there is a growing trend in the UK towards a practice already common in many other countries of basing home sales royalties on the *price actually received by the publisher.* This may be less than 50% of the retail price and is usually at least 30% less than that retail price. A fixed retail price for a book is a notion that only applies to the UK and certain other countries, where there are agreements — the one in the UK is called the net book agreement — between publishers and the book trade; these specify that books will be sold only at a price recommended by the publisher. Such agreements either do not exist, or are illegal, in many countries. If the publisher does not stipulate the royalty base you should ask. The base for export sales is usually *price actually received by the publisher*, although a few UK houses base their royalties for exports on the home retail price; if they do the rate is usually lower than the 'home sales' rate because overseas sales often involve bigger discounts, or extra handling costs for shipment or at the publisher's overseas offices, or payments to agents. Higher rates are often paid on sales exceeding a specified number of copies, often over 10 000 copies for a book aimed at a wide market, less for a specialist work. Before you even get to considering these details you should be happy about the publisher's marketing arrangements being right for your book.

At this stage you should try and assess your likely income. Ask for an estimate of sales if you did not get one in earlier discussions about marketing. For a specialist academic book this may be less than 1000 copies. Even if your book is targeted at a more popular market, or is a student text, expected sales may only be between 2500 and 10 000, depending on the subject matter, your target readership and the likely export sales. Ask also what the approximate retail price of the book will be. You need this to work out your likely earnings. For a lucky few the estimate will prove unduly pessimistic, for others it will be optimistic, for most it will be fairly realistic if your publisher is experienced in selling to the market you are trying to reach.

Having assessed the likely financial implications of an offer, you may express interest. If you are happy, do this, but emphasize that acceptance will be subject to a satisfactory contract.

In due course a draft contract will arrive; this is usually a formidable printed document. Modern word-processing and desktop printers give most draft contracts a 'final' look. Remember it is a draft; it can be, and probably should be, altered before you sign it. A contract is a legal document setting out the rights and responsibilities of author(s) and publisher. It will spell out fully the basis of payment for royalties and advances, the conditions under which either party may terminate the contract, and detail the penalties each may face for breaches.

It should specify precisely what rights you are assigning to the publisher and what benefits you may receive from sale by the publisher of subsidiary rights (including translation, radio or TV, electronic or mechanical reproduction, sales to book clubs and perhaps even comic strip rights; this last an unlikely source of revenue from a scientific or technical book, but you never know). For scientific and technical writing the most important subsidiary rights may be those for translation and for electronic reproduction; the first because translations sometimes bring in as much, or more, than English language sales and the second, electronic reproduction, because it is a fast growing area that includes reproduction on both computer discs, as CD-ROM, and possible multimedia developments; fields in which libraries, especially academic and other specialist libraries, are showing increasing interest. Indeed, it has been suggested that by the year 2000 many publishers of academic, scientific and technical books will derive as much as one quarter of their income from various forms of 'electronic' books. In the UK, the Society of Authors now publishes twice yearly a magazine, *The Electronic Author*, devoted specifically to 'electronic' aspects of writing and publishing.

Film or television rights might be important for work that expounds new ideas likely to catch the public imagination, or for a book aimed at the general reader, if it adopts a fascinating approach. Your work could attract a producer making a *Horizon* or an *Equinox* programme — a 1993 *Equinox* programme was based on Roger Penrose's *The Emperor's New Mind*.

The contract is likely to grant a licence to the publisher giving sole rights to publish your book and should (but may not) stipulate that this be done within a specified time; the licence usually assigns to the publisher control over the disposal of subsidiary rights; for these the author will normally receive stated percentages of any monies received from their sale.

Some contracts specify assignment of copyright to the publisher. This is usually acceptable if you are contributing only a single chapter to a joint-author volume, or entries on specific topics to a dictionary or an encyclopedia. Assignment of copyright is sometimes sought for complete books by reputable publishers, e.g. Cambridge University Press, and the practice is growing in academic publishing. One argument advanced is that this makes it easier to combat illegal reproduction (piracy) of books in certain parts of the world. The parties best placed to fight this are the copyright holder and the holder of any right that is being infringed. If the author retains copyright, a joint approach by author and publisher is needed; arranging this is cumbersome and the damage is often done before the pirated edition can be suppressed. If the publisher holds the copyright, more rapid action can be taken. I have some sympathy with this argument for academic books or textbooks, but the danger for you, the author, is that you lose all control over your work. Under a licensing arrangement there is usually a clause restricting the licence to a fixed term and specifying that all rights revert to the author if the book goes out of print and the publisher refuses to reprint.

My advice is to resist assignment of copyright, but if a well-known and reputable publisher requests this on grounds that it is easier to control piracy, then accept, but demand a clause in the contract specifying reassignment of copyright to you should the publisher's edition go out of print. The legal implications are important, so if in doubt seek professional advice. Never, in the case of a book, assign copyright for a flat fee in lieu of royalties. If you do, and your work becomes a best-seller, you will not get one penny more.

The position is different if you are commissioned to write an instruction manual describing how to use computer software or operate a piece of equipment. You will then almost certainly be offered a fixed fee and asked to assign copyright to the party commissioning you to write the manual. The fee may be governed by the amount of work, including background research or learning about the product, that you will have to do, but at 1994 rates in the UK it should be not less than £100 per thousand words, and at least double that if much preliminary work is required, or if you have to provide illustrations. In the US the dollar equivalent rates are often appreciably higher. Commissioned work of this sort is a profitable type of writing.

Other common clauses in contracts for books require the author to:

- prepare, or pay for the preparation of, an index, if one is needed;
- obtain permission, and pay any required fee, for using copyright material;

- guarantee that the work contains nothing libellous, obscene or blasphemous (Appendix A).

Some conditions in a contract may seem irrelevant to you, but most are there for a good reason; usually that the publisher has, or knows another publisher who has, had fingers burnt over these matters. A condition that would be irrelevant in a contract for a crime novel may be important for a textbook; e.g., provision for a financial penalty if an author fails to produce a new revised edition if the publisher deems it necessary.

Check how often royalties are to be paid; usually they are paid six-monthly, but some publishers only pay them annually. Because of the 'sale-or-return' trading arrangements between publishers and book-sellers, many publishers include a clause withholding, for two years, a fixed percentage (sometimes as high as 20%) of the first royalty payment to cover possible returns. This clause is common (and often relevant) for mass-market paperbacks.

Some publishers include a clause giving them a first option over an author's next work. Most will delete this if asked. Ask for deletion; if you develop a good relationship with a publisher and your next work is appropriate, you will almost certainly offer it first to that firm. If your next book is inappropriate to that publisher, or if you have a bad experience with a publisher, the clause is at best a time-waster; at worst a delaying mechanism, or even a potential barrier, to getting your next book published.

Get advice about any contract before you sign it. Friends who are established authors may help. These are often the best available source of advice, but that advice may fall short of perfection because contracts differ widely between publishers and also vary according to the type of book; comic strip rights are seldom important to a writer of research monographs, but they may be a main source of income for a writer of children's fiction. Legal advice is expensive and best sought from a solicitor who specializes in dealing with publishers' contracts. Any solicitor can tell you what each clause in a contract means, but one who is not a specialist may not know that '40% of the publisher's receipts from the sale of translation rights' is a poor offer. Sadly, advice from solicitors specializing in publishing contracts is usually so expensive that it might swallow up all the royalties for a part-time writer in a specialized subject area. In the UK the best value-for-money advice is probably that given by the Society of Authors to its members. This advice is free, in the sense that it is covered by a member's subscription. A snag is that the Society is a professional body and to join it you must be an author. You become eligible when you receive a firm offer-to-publish from a recognized publisher. Before this stage, that writer friend or colleague is probably your best source of help. The Society has negotiated minimum terms agreements (MTAs) with some publishers; unfortunately these seldom cover educational, technical or academic books. An outline of MTA provisions is given in the UK *The Writer's Handbook* (FRII). The Society also publishes a leaflet advising authors on points to watch when negotiating contracts, but not all these points are relevant to specialist writing. It has a group that deals specifically with matters of interest to scientific and technical writers, and corresponding groups for medical and for educational writers. Authors' associations in some other countries (section 10.4) also advise their members about contracts. Michael Legat, in *An Author's Guide to Publishing* (FRIII), discusses publishers' contracts in detail.

How much bargaining you can do with a publisher about royalty rates, advance payments, payments for subsidiary rights and other conditions depends on circumstances. For academic or specialist books do not expect a large advance; no advance is common. For a 'popular' science book sold to a mainstream general or paperback house consider yourself lucky to be offered a four-figure advance; any advances are commonly paid, one-third on signature of the contract, one-third on delivery of an approved manuscript, one-third on publication. Royalty rates of 10% of the retail price for home sales and 15% of publisher's net receipts for export are usually regarded as reasonable for hardback books; corresponding rates for paperback are 7.5% and 10–12.5%. If other bases are used for calculating royalties the percentages should be adjusted to something approximately equivalent. These are general rules; there are exceptions for books catering for special, and often very competitive markets, including school textbooks, where rates tend to be poorer rather than better. For several well-known series of books (including some on self-instruction and some that cater for named professional qualification examinations) the publishers have standard terms and conditions and will not alter these unless circumstances are exceptional; they will seldom do so for first-time authors.

If you submit CRC or copy for typesetting from disc you are carrying out part of the printing process. A publisher should, and reputable publishers who ask or encourage you to submit in this way usually do, pay you a fee for this.

When you are satisfied with the, often amended, draft contract and you have agreed any alterations with your publisher and signed, you must get down to serious writing.

If you feel bargaining over contract details is too awesome there is a potential escape route; become the client of a literary agent — if you can find one willing to act for you.

4.3.2 Literary agents

Literary agents are professional middle-persons (ugly, but politically correct language) between publishers and authors. The author is a client of the agent; the latter makes a living by deducting a percentage (usually 10–15%) from all monies earned by that author from work handled by the agent; so it is in the agent's interest to optimize these earnings. Do not spurn an agent on the grounds that there is no point in paying this commission, because a good agent can probably get better terms from a publisher than you could get by direct negotiation. In particular, the agent will probably handle subsidiary rights rather than assigning these to the original publisher, and will market these with vigour (which some, but not all, publishers also do), so where is the catch?

Unless you are aiming at the popular end of the scientific and technical market many agents will not want to know you. *The Writers' and Artists' Yearbook* and *The Writer's Handbook* both contain lists of literary agents, as do their USA counterparts. The entries bristle with terms like *No academic or technical*. Agents who do handle such books are only likely to be interested in established full-time authors. Even at the popular end of the scientific and technical market, an agent is unlikely to accept you as a client unless you are already making a *regular* income from writing of at least £5000, probably £10 000, per annum.

This is why an agent is even harder to find than a publisher. Nevertheless, if you have a steady writing income of £5000 or more per annum, you should seek an agent; a good one might boost that income appreciably.

4.3.3 After the contract is signed

After a contract is signed (you will have to sign a contract even if it is negotiated by an agent) keep writing steadily. Your deadline for submission, usually six months to one year ahead, may seem a long way off, but as you proceed you will experience some or all of the problems of writer's block (long periods when you can get nothing worthwhile on paper, no matter how hard you try), delays while you research essential material, unhappiness with some of your work (a sure sign that it needs revision), outside pressures that distract you from writing, a new discovery or development in your field that forces you to update or revise substantial parts of what you have already written. You will need also to leave ample time to check carefully the draft manuscript, hunting out those spelling inconsistencies and missing references, checking you facts and seeing that you have included all relevant material.

You must also check carefully your cross-referencing and your numbering of sections, tables and diagrams at this stage.

Later chapters in this book cover many practical points — how to explain, or avoid, jargon that might annoy various types of reader; preparing and using several kinds of tables and the many different sorts of graphs; how to select examples and exercises, compile an index, prepare CRC, manuscripts on disc, and so on.

If, at any stage you see problems looming that raise doubts about meeting the deadline for submitting your final manuscript let your publisher's commissioning editor know immediately. Publishers work to tight schedules that are planned many months in advance; they have to dovetail various aspects of producing your book with plans for others. Not only must printing schedules be arranged, but there are matters like jacket design, availability of copy-editors and designers to check manuscripts and mark them up for the printer (much of this is often done by freelances) as well as marketing arrangements to be made. A slip in a planned publication date might be disastrous for sales. For example, a textbook published six months before the start of the school or college year may be widely adopted for that coming year. If publication is delayed, not only will it miss out for that year, but its future may be blighted because a competitor gets in ahead of you for that next session. This commonly happens for books in rapidly developing technologies. Delay may close a narrow window of opportunity.

4.4 SELF-PUBLISHING

If you fail to interest a publisher, you might consider publishing your book yourself. Indeed for a highly specialized work with a sales potential of only 100 copies, this might be the only way to get published. Self-publishing is a growth industry. So-called *desktop publishing* software enables an author to produce attractively laid out material ready for printing either from disc or as CRC. Preparation in these modes is discussed in sections 5.2.2 and 5.2.3, and there is more about CRC in Appendix B. As I indicate in those sections you need to know something about book design and layout to use these techniques satisfactorily. However, that is probably the easiest part of self-publishing. Printing and binding will almost certainly have to be contracted out (unless your organization has a printing unit) and you will either have to place marketing in professional hands (an expensive business) or do your own — the latter both time-consuming and potentially expensive.

Journalist Keith Austin self-published his comedy thriller *Taking Liberties* after many publishers had turned it down. He did his own marketing and even managed to land an order for 3000 copies from Britain's largest book-selling chain, W.H. Smith; an achievement that would, if equalled, put a 'cat-that-ate-the-canary' grin on many a publisher's face. His is a somewhat isolated example of great success, and self-publishing probably has it best potential in books likely to appeal to a highly specialized readership, for then marketing can be targeted at that group. If you are an expert in the field you will probably know how to do this; for example, nearly all potential readers might be members of one or two professional societies or trade associations and it might be relatively easy to approach members by mailshot or by advertisements in those organizations' publications.

The *Guide to Self-Publishing* by Harry Mulholland (FRIII), himself a self-publisher, gives practical advice to would-be self-publishers, as does another self-publisher, Peter Finch, in *How to Publish Yourself* (FRIII). A number of books on the USA market (again many of them self-published) where self-publishing has an even longer history, and perhaps even greater potential than it does in the UK, also give sound practical advice.

You must not confuse self-publishing with vanity publishing. Vanity publishers are firms, or individuals, who offer to perform for a fee (usually a hefty one) all the functions of publishing a book such as printing and marketing. Their efforts seldom live up to their promises. Most vanity publishers are well-known to, and may be blacklisted by, the book trade. Read the annual warning about vanity publishing in the *Writers' and Artists' Yearbook* (it is on p. 267 in the 1994 edition).

5

The book publishing process

5.1 THE PUBLISHER'S WORLD

You will deal with a commissioning editor while negotiating a contract and throughout the writing period — this person may also be called a senior, or an executive editor, or just an editor, in some publishing houses. After the manuscript is submitted and approved, your main contact is likely to be a sub-editor, or a production editor, who continues as your key link during the rest of the publishing process, although at various stages you may be contacted by a copy-editor, a representative of the marketing department and perhaps also by a designer and by a proof-reader or by someone with a fancy name like schedule co-ordinator. Your sub-editor will pass your manuscript to a copy-editor, often a freelance familiar with the subject you are writing about. The copy-editor is expected to pick up not only spelling mistakes and punctuation and grammatical errors, but also to watch out for obscurities, ambiguities and factual errors; also to check that your manuscript conforms with house-style. To learn the details of a copy-editor's job and how it interfaces with writing, read Judith Butcher's *Copy-Editing* (FRI). I say more about the relationship between copy-editor and author in section 10.1.

Except for clear mistakes, obvious grammatical errors, or corrections that are trivial but definite improvements, any changes proposed by a copy-editor should be referred as 'queries' to the author. This may be done directly, or through the sub-editor. Depending on the subject matter, there may be a hundred or more queries and suggestions even for a well-prepared manuscript. Many of the points may be closely related and intended only to bring your text into line with house-style. Other queries may be about inconsistencies in spelling or punctuation; mismatches between citations in the text and those listed in the references (e.g. different publication dates or different spellings of an author's name); reordering of words or sentences; possible obscurities or ambiguities. The copy-editor may suggest that some concepts or jargon need further explanation, or point out that you have used an expression on p.6, but not defined it until p.29, the sort of trap all writers fall into.

Good copy-editing is a demanding task. The Spon guide for copy-editors points out that

> *If copy-editors do a good job no-one notices! If they do a bad job, everyone does!*

Some queries may irritate you, even seem trivial, but consider each carefully; if you disagree, say so, giving a reasoned argument for your stance. A copy-editor once suggested I change *sulphate* to *sulfate*, because this (p. 7) is now a preferred spelling by chemists in the UK, as well as the recognized continental and American spelling (markets where it was hoped the book would sell). I counter-argued that (i) I was quoting a worker who used the spelling sulphate and (ii) most interested readers of that example would be biologists or agronomists who, in the UK at any rate, prefer sulphate. I won, but I would not have lost sleep if pressed to surrender on so minor a point. I give this illustration only to show how easily queries sometimes arise. Some authors (and publishers) are touchy about such detail. I say more about the nature of queries and responses in section 10.1, and about an author's right to resist 'unreasonable' editorial changes, in Appendix A, section A1.3.

Copy-editors do some marking of a manuscript for the printer, but final decisions on layout are usually made by a production editor or controller, or by the publisher's design or production department. They, too, will often want to discuss some points with an author; for example, jacket design of your book. If your book is to be published in a series this may follow a standard format; if not, ask to see the design — some cover-artists gets carried away. It is an open question how far jacket design influences sales. Suitable eye-catching covers certainly help, but even if it is an eye-catcher, a subtle or obscure design may put off potential readers with a fear that the book is too erudite for them. I think this happened to one of mine, although both my commissioning editor and I liked the cover. For another book I thought the cover revolting, but it was certainly eye-catching and the work sold well; for yet another the cover was eye-catching (and I thought appropriate), but the book only sold moderately well. All this I found confusing, except as an indication of the obvious truth that success depends upon more than jacket design. However, publishers consider cover design important, and there is ample evidence that it often influences sales.

The marketing department may approach you before you submit your final manuscript; if not, it should do so soon after. If you are asked to fill in an, often long, questionnaire do this carefully and give the matter plenty of thought. Constructive responses could double sales. In particular, answer questions about target readership; highlight any professional, trade or academic groups (e.g. the pharmaceutical industry, civil engineers, geologists, systems analysts, quality control supervisors, business executives, the insurance industry) to whom your book may be of particular interest. Suggest journals likely to review the book if sent a copy and, if asked, list professional societies or trade organizations whose members may be interested. Do not expect leading daily newspapers to review your masterpiece unless it is aimed firmly at a mass readership. The questionnaire may ask you to supply information for use as a basis for the blurb on the jacket. Whether or not there is such a question, ask your sub-editor to let you see the copy for the blurb. You may not like what you see, but remember it is essentially advertising copy, so it should flatter your efforts. If you have an established reputation in your field, do not be bashful about being described as 'a world authority'. However, draw the attention of the sub-editor to factual errors, or to any serious omissions.

Unless your book is prepared as CRC (section 5.2.3 and Appendix B), in which case you must proof-read before final submission, you will have to proof-read the typeset material. Your sub-editor will probably be your link for this important task, described more fully in section 5.4, but you may also be contacted by the publisher's own proof-reader.

The publication process will be streamlined if, at all stages of writing, you present your material in a correct format, complying with standard conventions or instructions from your publisher.

5.2 PREPARING YOUR MANUSCRIPT

In this section I deal first with books prepared as conventional typewritten or word-processor manuscripts; then with books submitted on disc for direct typesetting and finally with CRC. In each case, precisely what you should do depends upon constraints imposed by your publisher, and upon guidance given on house-style, preferred conventions, etc., so I can only comment on general aspects. There are certain topics, e.g. using abbreviations, and how to write numbers, that raise specific problems and I consider these in section 5.3.1 and 5.3.2, while section 5.4 deals with proof-reading, and section 5.5 with preparing an index.

5.2.1 Preparing a conventional manuscript

You will enhance your reputation, and probably speed publication, if you submit a manuscript that complies with requirements set down in any general instructions for authors provided by your publisher and with additional stipulations specific to your book, e.g. you may have been asked to use the Harvard system (section 2.2.5) for references and also to include *further reading lists* at the end of each chapter. If you are not given a style guide, or if there are points the one you are given does not cover, the following recommendations should help.

- Your manuscript should be presented in a series of *bundles*. The major *bundle* consists of the main text. Subsidiary, but very important, additional bundles will contain tables, figures and perhaps plates or computer listings.
- The main text should be double-spaced and on European A4 or American letter-size paper with adequate margins at top, bottom and each side (section 4.1.2). Use one side only of each sheet. Commence each chapter on a new page (manuscript pages are called folios in the publishing business) but number pages (folios) consecutively at the top right; i.e. if Chapter 6 ends on p.72, then p. 73 is the first page of Chapter 7. If you have at some stage to insert additional material between, say p. 16 and p. 17, number pages with the new material as 16a, 16b, 16c, ... At the foot of page 16 put *page (or folio) 16a follows* and at the foot of the last inserted page put *page (or folio) 17 follows*.
- Use one font (preferably either 12-point Times Roman or Courier) throughout for text and also for chapter and section headings, each of which should be on a separate line with its status (chapter heading, section heading or subsection heading) made clear by differing indentation or by a pencilled marginal note.
- Left justify the text (i.e. leave right margins ragged). Do not insert extra space between paragraphs, but indent the first line of each paragraph, except the first in each section, by 2 or 3 characters. Except for the last page of each chapter, every page should, if possible, contain the same number of text lines, counting a section or other heading as a 'text line' in this context.
- Chemical equations and mathematical formulae, unless very brief, e.g. $I=Prt/100$, should usually be displayed, e.g.

$$2CuSO_4 + 4KI = 2CuI + I_2 + 2K_2SO_4$$

- Avoid over-use of *italic* or **bold**, but use these if they are essential (e.g. in botanical nomenclature use italic for genera and species, or in mathematics use italic or bold according to convention for variables, vectors, matrices, etc.). Unless your word-processing facilities will produce clear *italic* or **bold**, indicate essential use of these by single straight or by single wavy underlines respectively.
- Tables, diagrams, plates, etc., should each be presented in separate bundles, but the position of these features in the main text should be shown by appropriate indicators (each displayed on a separate line or pencilled in the margin at the appropriate point) taking the form

[Table 7.1 about here]

or

[Figure 9.7 about here]

Omit the word *about* if exact positioning of a table or figure is essential. Tables should be collected in one bundle, each being numbered with a full descriptive heading. An exception is usually made for tables occurring in exercises. These are not numbered, and if they are very short they may generally be included in the text. Otherwise, use a separate sheet for each table and follow the guidelines on layout given in section 6.2. Figures should be collected together in a further separate bundle, the number of each figure being indicated in pencil on the back close to one corner. Each figure should be on a separate sheet and accord with specifications in section 6.3. List figure captions, numbered in order, on a separate sheet or sheets forming a *captions* bundle. If plates (e.g. half-tone or colour photographs) are included these should be in yet another bundle, with a separate list of captions following specifications given in section 6.4.

- Some publishers prefer preliminaries (calling them 'prelims') — list of contents, preface, acknowledgements, or, if you prefer, acknowledgments (but stick to one or other spelling), to form a separate bundle — but if this is not indicated, include this material at the start of your main bundle numbering the first page (folio) p.1.
- Appendices (if any) should be included in the main text bundle, following the last chapter. References are usually placed next. If exercises are included, and answers to these are given, these answers should appear after references but before the index (which will be prepared later). Sometimes, especially in mathematics, statistics or engineering books, there will be general reference tables, e.g. of the normal probability integral, or of significance levels for the t-test, or computer program

listings. These usually appear after *references*, or sometimes as an appendix, and unless otherwise instructed, may be included in the main text bundle or in the tables bundle. Many publishers reproduce computer listings as camera ready screen print-outs, rather than typesetting by re-keying, as this reduces possible confusion between '0' (zero) and 'O' (capital letter) or between '1' (figure) and 'l' (lower case letter), not to mention typesetting errors that can make a listing useless.

.● Check carefully spelling, punctuation, grammar, clarity, consistency and other features of 'good' writing before posting your manuscript. In particular, if writing has extended over a long period, watch out for contradictions, ambiguities, or subtle changes in meaning between your earlier and your later writing.

● Your publisher will usually require two copies of the manuscript; a good photocopy generally suffices for the second.

After submitting your manuscript you may relax — or get on with all the other jobs you neglected while you were writing — until you receive a request for marketing information (if you have not had one earlier) or until the copy-editor's queries arrive. Deal with these efficiently and expeditiously. Your next job is proof-reading and preparation of an index, topics I deal with in sections 5.4 and 5.5.

5.2.2 Submission for typesetting from disc

If your publisher accepts, or encourages you to submit your manuscript on disc ready for typesetting without re-keying, you must be clear in advance about the required format. Publishers who allow this approach supply instructions which must be followed carefully. These generally list codes to be used to indicate features like chapter and section headings and sometimes additional attributes such as bold or italic. In a few cases they may include codes specifying all fonts, point sizes and other characteristics that will permit direct setting from disc without an interface through a typesetter (who would normally insert such design features), but it is seldom you will be asked to use these. There will also be instructions about how to present mathematical formulae, tables, diagrams, etc. There will probably be a list of 'do's' and 'don't's' covering matters like the use of shading, displaying formulae, justifying text, etc. A printout (hard copy) of the material on disc will be required for editing purposes.

The main problem with typesetting from disc is interfacing between word-processing software and hardware used by an author and the corresponding equipment used by the typesetter. To illustrate some relevant points, I consider some, but by no means all, of the instructions provided by Spon and Chapman & Hall for authors preparing manuscript on disc. Other publishers' instructions may differ, especially about the extent of coding, and the codes to be used, to indicate features.

The main advantage of typesetting from disc is that it avoids re-keying your text (a common source of errors). The down-side is that any keying errors you make will be perpetuated unless they are picked up by the copy-editor. Thus editorial accuracy and consistency are important. Your material on disc, and the accompanying hard copy, will be checked by a copy-editor who will, where appropriate, make alterations on disc; these may include both textual amendments agreed with the author and insertion of additional codes.

Most major computer hardware and word-processing, desktop publishing and graphics software packages are compatible with commonly used typesetting facilities, but always check with your publisher before starting whether this is so for your equipment. Keep back-up copies on disc and up-date these regularly. That, I expect you know, is something you must do with all computer generated material that is not of an ephemeral nature.

Potential interfacing problems arise with tables and diagrams, or with displayed material such as formulae and equations. Sometimes it is best for the typesetter to re-key such material, and if so it is usually required in separate files with clear indications of where it is to be inserted. A careful check must be made at the proof-reading stage to see that this material is accurately reproduced and inserted at the right place.

It saves time at the copy-editing stage if your manuscript complies with the publisher's instructions for house-style, coding and file structure. Usually separate files are required for each chapter, for appendices, for glossaries, for preliminary material (preface, contents list, acknowledgements, etc.), for references, for the index, as well as for tables and for figure captions. Files should be given informative names like Gipch1.txt, Gipapp.txt, where here I am using 'Gip' to code the title *Getting into Print*. The procedure with diagrams varies with circumstances; computer generated diagrams are often printed from disc, but some artwork may be supplied ready for camera reproduction.

Codes are needed to indicate various levels of heading and for special symbols (Greek letters, multiplication signs, etc.). Sometimes you may use your indigenous word-processor codes for special symbols, or you may specify your own codes; in the latter case you must supply a list of these for the typesetter. Commonly used codes for headings — inserted on the disc before the relevant heading, and sometimes repeated or followed by a different 'closing' code at the end — are <<CH>> for a chapter heading,

<<A>> for a main section heading and <<B>> for a subsection heading. For special symbols, Spon recommends the '&' code. For example, the multiplication sign, '×', might be coded (i.e. typed on your keyboard at the appropriate spot) as '&x'. Similarly, the lower case Greek beta, i.e. 'β', might be coded '&b'. With the standard typewriter or computer keyboard, there is often no opening quote (') and a special code may be needed for this; another potential problem is distinguishing between the hyphen '-', the en-rule '–', the em-rule or dash '—' and the minus sign '−'. A commonly used coding device (not only on disc, but also in typed manuscript presentation where there might be ambiguity) is to use '--' as a code for a dash or for an en-rule to separate numbers. If you do this, leave a space on each side if this is required in print; e.g. if you want the typeset version to read:

John — a young devil by nature — quickly jumped aboard

you should type

John -- a young devil by nature -- quickly jumped aboard

but for *pp. 17–21* type *pp. 17--21* **not** *pp. 17 -- 21*, which would reproduce as *pp. 17 — 21*. There is an increasing tendency for printers to make no distinction between an em-rule and the en-rule and to use the latter both for a dash and for a divider between numbers.

Because requirements for material on disc differ between publishers, I have only described the rudiments to indicate some potential benefits and pitfalls. Improving technology means that in future authors are likely to be able to do more of their own typesetting. For example, one of the main jobs for the typesetter working from disc is to insert design features (different fonts, different spacing between words and letters and other features) that give the printed book a 'professional' look. Facilities to do this sort of thing are available in most modern word-processing and page-making packages. Publishers may not encourage authors to include these features because most writers lack the design skills that enable the trained typesetter to use them effectively.

5.2.3 Submitting camera ready copy

Many publishers have reservations about accepting CRC from authors. This may be attributed to (1) the widespread use of CRC for collections of conference papers, where the aim has been to speed publication with, until recently, too little regard for design quality and (2) the difficulties faced by authors due to software and hardware limitations and their sometimes knowing little about design features. The first reservation is justified. Many conference proceedings using CRC have involved many authors using different software and hardware, a variety of fonts and point sizes, margins of different widths, some using full justification, others not, and variable numbers of lines per page — not to mention differing conventions in citing references and in the use of section and subsection headings. This has happened even when publishers have issued quite detailed instructions, largely because some authors do not have the facilities to comply with all of them, and others are too lazy to do so. It is also true that many authors have little knowledge of, or interest in, book design. The minimum requirements for producing satisfactory CRC are:

- an up-to-date version of a major word-processing package or desktop publishing pagemaking software;
- a modern inkjet printer or a laser printer with enhanced 300 dpi (dots per inch), or preferably 600 dpi, resolution;
- an understanding of the rudiments of book design, or access to a publisher's guide to requirements for CRC.

This book is produced from CRC. It is the third book I have prepared in this way; with each I have learnt from my mistakes. I have regularly updated my software and hardware to improve the quality. I enjoy producing CRC because I am interested in — but not an expert at — book design. I use appropriate software and a laser printer with 600 dpi resolution. Even this resolution is not as good as that with machine typesetting. To overcome this problem I use a 14-point Times New Roman font for basic text, which, after reduction to 80%, will appear as though it were approximately an 11-point font. I use a smaller point size for tables and for captions to diagrams. My software enables me to produce mathematical symbols, Greek letters, opening quotes ('), en-dashes and em-dashes and minus signs (as distinct from hyphens), and to control spacing between letters and words. For example, in writing a 5-digit number most publishers like a thin space between the second and third digit. Using standard spacing gives a form 25 413 when 25 413 is preferred. You may need a sharp eye to spot such differences.

Kerning is a useful feature to give a more professional typesetting appearance. This adjusts spacing between specific letter combinations, bringing, for example, the letters in a combination like WAV closer together than those in GWZ; this reflects the opposite 'matching' slopes in adjacent letters in the former combination and makes for smoother reading. Many word-processing packages include an automatic kerning facility; but this may give bizarre results with certain fonts when used with some computer and printer combinations.

Other problems with CRC are controlling alignment at the top and bottom of pages; incorporating tables, figures, etc.; getting appropriate spacing between section headings and text; dealing with footnotes; providing headers in appropriate form. All are relatively easy to overcome with modern software, but you should follow a publisher's house-style or other guidance about use of fonts. Most publishers' design departments are happy to comment constructively on a few pages of 'specimen' CRC. Particular care is needed to ensure consistency in reference lists, etc. Publishers will want your so-called 'final' version for copy-editing and you will then have to adjust this to meet house requirements. This may mean using different fonts or point sizes for certain material (e.g. tables, displayed formula); such changes may affect large segments of CRC, so it pays to get these things right in your original submission. In diagrams there are potential problems with shading; it is often best for your publisher's design or production department to add this in appropriate form using screens. Discuss this with your publisher early in the process. I elaborate on many of these points with specific reference to preparing the CRC for this book in Appendix B.

5.3 SOME PROBLEM AREAS

I could have discussed abbreviations and ways of expressing numbers in Chapter 3, for clearly they are aspects of microstyle. I decided to consider them here because they are topics that crop up as one writes, and it is often difficult to decide policy about them in advance.

5.3.1 Abbreviations and units

Should one write *Dr Smith* or *Dr. Smith*? *CRC* or *C.R.C.*? *USA* or *U.S.A.*? In these three cases my preference is for the first option, but your publisher's house-style may specify the alternative. Unless an abbreviation is well known, spell it out in full the first time you use it, putting the abbreviation immediately afterwards in parentheses, e.g. camera ready copy (CRC), or if it seems more appropriate reverse that order as I did on p. 75 with *dpi*. Thereafter use the abbreviation only, except perhaps if it has not been used for many pages, or for several chapters, when there may be a case for repeating the full version and abbreviation. I did that with CRC in this book. There is no need to spell out well-known abbreviations such as USA or UK, or a few abbreviations that are better known than the full terminology, e.g. AIDS and DNA. Indeed, some acronyms such as RADAR are now often regarded as words in their own right; we usually simply write 'radar'; many people will not even know that it is an acronym for *radio detection and ranging*. There are borderline situations; people in computer circles know what ASCII means, but those outside may not know that it is the *American Standard Code for Information Interchange*.

Use of a full point after an abbreviation is decreasing. The Spon copy-editors' guide recommends a full point only if the final letter of an abbreviation is not the final letter of the word being abbreviated, e.g. use *Fig.* for *Figure*, but *Dr* rather than *Dr.* for *Doctor*. Full points are always used in *e.g.*, in *i.e.*, after *etc.* and after the *al* in *et al.* Whether to write *6 pm* or *6 p.m.* may depend on house-style.

Abbreviated species names such as *E. coli* require the full point. However, the trend is to omit the full point in sets of initials that are essentially acronyms, e.g. USA, DNA (not U.S.A., D.N.A.).

International scientific (SI) units, e.g. *kg*, generally do not take the full point as these are regarded as names of units; however abbreviations of imperial units (still widely used in the USA) such as *in.* for *inch* are usually classed as abbreviations and are followed by the full point.

The names for SI units are standard and should be adhered to (including the use of capitals where appropriate, and the omission of full points), e.g. *g* not *gm* or *gm.* or even *g.; km* not *Km*, but *J* not *j* for joule; *l* not *L* or *lit* for litre. I have written correct and incorrect versions in italic to highlight them; *italic* is not used for these names in text. Thus, write 16 mg/l or 16 mg l^{-1}, but use one or the other consistently, rather than 16 mg per l. However, in section 2.2.4 I used the expressions *per annum* and *per km* in text and in a table footnote in a situation where that form is often used. Broadly speaking, *mg/l* is preferred in the life sciences and in medicine and *mg l^{-1}* by physical scientists, mathematicians and engineers. Note the space between *mg* and *l^{-1}*.

You may feel, as I do, not completely at ease with the above guidelines for punctuation in abbreviations (which may differ between publishers), but try to be consistent. If you are in doubt, I recommend you follow Hart's rules.

5.3.2 Numbers

I was trained as a mathematician, so numbers should be second nature to me, but I often run into trouble presenting them in print. When numbers are used as measurements in specified units there is little problem; use arabic numerals, e.g. 7 g, 23 m, 16 in., 7.25 kg, 0.423 mm. In the final example note the leading zero before the decimal point for a number less than 1; always include this; there is a danger that a 'leading' decimal point without a preceding zero may get 'lost' in reading, or even in typesetting.

Presenting numbers that represent counts is more difficult. Publishers usually recommend spelling out numbers from one to ten. For example,

There were two bananas on the shelf

but

There were 23 oranges in the box.

This leaves an unhappy situation with a mix of numbers above and below ten. I do not like

John has six oranges, 13 bananas, five pineapples and 17 plums.

Unless your publisher's house-style demands otherwise, I recommend arabic numerals throughout in that example.

Never start a sentence with a numeral; either spell it out or restructure the sentence; e.g. not

14 members were present

but either

Fourteen members were present

or

There were 14 members present.

In some countries the comma is used as a decimal point, e.g. 25.417 is written 25,417. So avoid the comma as a group separator and use instead spaces, writing 25 417 209 in preference to 25,417,209. Usually no separator is used in integers between 1000 and 9999. Some very large numbers, e.g. 12 500 000 000 may be better written in *standard* or *mathematical* form as 1.25×10^{10}. Avoid writing this as 12.5 billion, because there is confusion as to whether *billion* is one thousand million or one million million, although the former interpretation is now more common on both sides of the Atlantic.

In tables with numbers in columns it is sometimes appropriate to align on the decimal point and sometimes to right align on the final digit. Whatever is decided, do this consistently.

A problem often overlooked is the occurrence of neighbouring numbers that may be run together in reading, e.g. *11 100-day-old progeny* might be misread as *11100-day-old progeny*, or if the first hyphen is omitted, as *11100 day-old progeny*. The best solution is probably to rephrase, e.g. *11 progeny, each 100 days old.*

Occasionally, as there is with spelling, there are slight differences between British, continental and American use of numbers; I have already referred to using a comma rather than a full point for the decimal point in some countries, and the billion difficulty. In expressing time in the USA it is common to use *6:17 pm* as an alternative to *6.17 pm*. Remember also that while *no.* is used as an abbreviation for *number* in the UK, a common American abbreviation is *#*.

Avoid the abbreviations ', " for feet and inches, or for minutes and seconds in angular measure. When using abbreviations for chemical elements or for units be aware of possible ambiguity like that between *C* for the element carbon and *C* for degrees Celsius.

I dealt with the correct abbreviated names of units in the SI system in section 5.3.1. If imperial units are used the unit name is singular, e.g. a *12-inch ruler*, or a *3-mile race*, or a *5-lb. weight* (not *5 lbs. weight*). However, *I passed him 3 miles down the road* is correct, for here *3* is a 'count' of three one-mile units. If you are converting measurements to different units, e.g. imperial to SI, do not give an impression of spurious accuracy by including too many decimal places, e.g. write *7.23 in. (183.6 mm)* not *7.23 in. (183.642 mm)*. Be careful also to get the decimal point in the right place. There could be serious consequences in a medical context, for example, if a mistake of this kind occurred in converting a drug dose.

5.4 PROOF-READING

For CRC your proof-reading is done before you submit your final manuscript. However, as I pointed out in section 5.2.3, your CRC will be copy-edited and returned to you together with queries; on many of these you may have to take action. When doing so be very careful to avoid introducing further errors. Completely proof-read your corrected copy! With most word-processing packages, when one replaces a section of text by substitute material it is all too easy to forget to delete that old text! If alterations to the text affect pagination you will have to submit new CRC for all amended pages. The addition of two or three words sometimes affects pagination to the end of a chapter, or even, horror of horrors, for the remainder of the book. When major changes of this sort are made problems can arise with headers if sections are moved to a new page.

If you have submitted a conventional manuscript, or discs for direct typesetting, you will be sent proofs (sometimes two copies, one of which you keep) for correction. If you receive two copies make your corrections on both, so that you can refer to your copy if any corrections are queried by the publisher. Sometimes the publisher's own corrections will already have been made on the copies sent to you; there may also be typesetter's or publisher's queries that you must answer. You may receive instructions to mark typographical errors in one colour and any changes you make to text in a different colour. There are conventional marks for indicating necessary changes — insertions, deletions, transpositions, changes to capital, lower case, bold, italic or different fonts or point sizes. Your publisher may send you a list of these marks, but there is a full list in *Hart's rules* (FRII) and in many other reference books for authors and publishers including O'Connor (FRI). There are minor differences between standard UK and USA markings.

The prime task at the proof-correction stage is to correct typographical errors or make **essential** editorial amendments. It is not an opportunity to revise your manuscript. Indeed, if you attempt to do so your publisher will either disallow the changes, or will send you a bill (which may run into hundreds, even thousands, of pounds or dollars) for the cost of making the changes. Certain minor amendments may be allowed; for example, if important new work has been done on a topic it may be possible to insert a reference to this. Try to do so in a way that will minimize text resetting, either by deleting some passage of exactly the same length at the point of insertion, or by putting the reference in a paragraph that has only a partial text line at the end, which will merely be filled by the new reference.

Ideally, proof correction is a two-person job. Get someone to read from your original manuscript while you check that the typeset copy agrees. Even this will not suffice, you must re-read the typeset material a word at a time, checking each word for spelling or typographical errors. It is all too easy to pass misspellings such as *neccessary, accomodate*, even if you had the correct spelling in your manuscript. It is also easy to overlook a *were* that should be *where*, a *there* that should be *their,* an *its* that should be *it's,* or vice versa.

Checking references is an important, almost soul destroying, task. You and the copy-editor should have checked at an earlier stage the correspondence between references cited in the text and those listed (section 2.2.5) and picked up inconsistencies such as a journal title sometimes being given in full, sometimes abbreviated; sometimes only first page numbers being given for papers, at others both first and last page numbers, but a few slips like this may have been missed. Also, the typesetting of references is made difficult by the use of **bold** and *italic* in certain conventional ways and it is all too easy for setting mistakes to be made in dates, volume numbers and in first and last page numbers. Ideally, if time permits, each reference should be checked against the original publication, but with a long list of references

— there were more than 200 in my last book — that is a daunting task and time schedules may make it impossible.

Unless you prepared CRC (when the index is submitted with your original material), it is at the proof stage that you must prepare the index and also insert correct page cross-references if these have been included in your manuscript as p. 000. Preparing the index is covered in the next section.

Your corrected proofs will be checked against a set that has been read by the publisher's own proof-reader (who may well detect mistakes you missed); all alterations will be collated on one set of proofs. If you have asked for textual changes a decision will be made at this stage whether to allow these. In some cases you will be pressed to justify a proposed change; e.g. to indicate whether it is removing a previously undetected error, or clarifies an argument, or reports an important new development since the time you prepared your manuscript. There may be other queries; in one book where I gave solutions to exercises a proof-reader checked all my answers and (justifiably) queried several.

5.5 PREPARING AN INDEX

With CRC this is done before you submit your copy; in other cases it is done at the proof stage, usually under pressure to meet a deadline. Reduce that pressure by deciding, well before you are due to get the proofs, what you want to include in the index. On your copy of the manuscript mark with a highlighter each word or topic you want to enter; arrange these items alphabetically with appropriate cross-references in a dummy index in which you only have to insert the final page numbers (you may insert these immediately if you are preparing an index from the final version of CRC). Many word-processing programs have a facility for index preparation; these may suffer from certain practical limitations (especially in relation to cross-referencing) but they provide a useful starting point. You might also use an appropriate data base package, or a card-index system for your dummy index.

When deciding what to include in an index try and put yourself in the position of a potential reader. To what word or words is a reader likely to refer for information on a given topic? Generally, an entry will be a noun or a well-known combination of adjective and noun such as *brown bread* or *secondary reaction*.

Do not give a page reference to every mention of a topic, but list only a page or pages where there is substantial information. If there are more than about half a dozen references to one topic try and break these down under separate sub-headings. These may be preceded, if necessary, by a few page references of a general nature that do not fall under any particular sub-category. For example, in Sprent and Sprent (1990) the authors gave the following entry under *Actinorhizal plants*:

Actinorhizal plants 30, 55–63, 118, 172, 195, 205
   direct use in forests 132–5
   evolution of 205
   indirect use in forests 135–6
   infected cell structure 50–1
   nodulation in 58, 61
   oxygen relations in 62

The secondary level entries are mostly descriptive phrases and these are arranged in alphabetical order; this is customary, but it is usual to ignore prepositions or conjunctions in this ordering. Note the punctuation, and the way page numbers are presented in this example. Although the problem did not arise here, the convention is that if a reference to a topic is spread over two or more pages this is indicated by the form 142–3, but if the references on successive pages are independent write 142, 143.

Cross-referencing takes two forms. The *see also* form is put at the end of a reference to indicate another heading where related material may be found. The *see* form is used when two topics are so nearly equivalent that it is convenient to put all items under one entry and cross-reference to this from the other. For example, in statistics the terms *hypothesis test* and *significance test* mean virtually, though not quite, the same thing, so it makes sense to list the various tests (with appropriate subclassifications) under *hypothesis tests* and under *significance tests* to put

Significance tests, *see* Hypothesis tests

See is also used if there is doubt about where a reader might look for information. For example one might put information on secondary reactions under *Reactions, secondary*; it would then be appropriate to include an entry under *Secondary reactions* in the form:

Secondary reactions, *see* Reactions, secondary

See and *see also* are printed in italic.

An index is usually printed with two columns per page. If you are submitting CRC your index should have that format; most word-processing packages provide the necessary facility. If your index is to be typeset present it double spaced as you did your manuscript, with sub-headings indented.

Sometimes two indexes are given in a book — a subject index like that just described, and an author index. The latter may not be needed if there is a list of references and nearly all authors named in the text are associated with a citation. If a person's name is associated with a particular result, technique or piece of equipment it is in order to indicate this by an entry in the Subject index, e.g. *Pythagoras's Theorem, Student's t-test, Green's function, Krebs' cycle, Kjeldahl method, Gibbs' equation.*

6

Special tools

6.1 SCOPE

I described in Chapter 2 an incidental use of tables, graphs, examples, exercises, displayed formulae or equations, and instructions for experiments, to make layout more attractive. More importantly, these are characteristic, and often essential, features of scientific and technical writing, yet authors sometimes under-rate the extent to which they enhance a book, paper, report or instruction manual. They even look upon preparing tables or diagrams and selecting examples and exercises as unwelcome drudgery. Which features are important clearly depends on subject matter and target readership. Exercises for students are needed in most textbooks in mathematics and for many in physics, but are unlikely to have a place in a field handbook for geologists. Descriptions of experiments or techniques are usually appropriate in textbooks on practical aspects of chemistry, physics, biology and engineering. Diagrams are often vital explanatory tools in instruction manuals.

In this chapter I indicate some uses for, and consider the microstyle of, these devices. Because their selection and preparation often involves information retrieval, including the use of databases, I comment on the collection and collation of information in section 6.8

6.2 TABLES

The commonest type of table is a data table. Tables 2.1, 2.2 and 2.3 are examples. More generally, tables are used to summarize information, qualitative or quantitative, that can be arranged in categories. Keep tables simple; use one only if it is the best way to give information; alternatives are to use a graph or embody information directly in text.

Table 6.1 is a qualitative table that might be appropriate in a book describing the role of nitrogen in plant growth; it summarizes the ways that oxygen is involved with nitrogen-fixing organisms.

Table 6.1 Some ways in which oxygen is involved with nitrogen-fixing organisms

Effect	*Comment*
Repression of nitrogenase synthesis	Widespread
Inactivation of nitrogenase	Invariably, although sensitivity may vary
May be necessary for nitrogenase activity in some aerobes	*see* p. 725
May limit nitrogen fixation in some symbioses	Low O_2 essential for some processes
May be used in hydrogen oxidation	Is in most species studied

There is a table like this in Sprent and Sprent (1990). That book, like many others, also contains tables with a mix of numerical data and comments, as well as tables of numerical data only. A table must be linked to a description in the text, but it should also be as self-contained as possible. The heading should be informative, and when needed, row labels must be supplied on the left and column headings at the top. Horizontal rules are useful to break up sections of a table, but most publishers dislike vertical rules between columns. This dislike has historical roots connected with typesetting. That problem is largely overcome by computer-setting, but a more reasoned argument against vertical rules is that they give a table a cluttered appearance, and that appropriate use of tabs is more pleasing. This is a matter of opinion. Do you prefer Table 6.2 or Table 6.3?

I prefer Table 6.2. Both tables, often described as 'two-way' tables, show the same quantitative data categorized by dose level of a drug and by side-effect incidence. Sometimes two-way tables are used when the responses are not counts but are general algebraic forms or qualitative responses.

In Table 6.4 the responses are in an algebraic form representing probabilities.

Table 6.2 Numbers of cases where no side-effect or a side-effect of given severity was observed at three dose levels of test drug A75

Dose	Side-effects		
	None	Mild	Severe
100 mg	50	0	1
200 mg	32	3	0
300 mg	29	6	2

Table 6.3 Numbers of cases where no side-effect or a side-effect of given severity was observed at three dose levels of test drug A75

Dose	Side-effects		
	None	Mild	Severe
100 mg	50	0	1
200 mg	32	3	0
300 mg	29	6	2

Table 6.4 Probabilities for the four possible outcomes when two biased coins with probability of heads p_1, p_2 respectively are tossed

		First coin	
		Heads	Tails
Second coin	Heads	$p_1 p_2$	$(1-p_1)p_2$
	Tails	$p_1(1-p_2)$	$(1-p_1)(1-p_2)$

Table 6.4 is a valid table, but an alternative is to display the information in the main text like this:

If two biased coins with probabilities of heads p_1 and p_2 are tossed the probabilities of each of the four possible outcomes are:

Both heads	$p_1 p_2$
First heads, second tails	$p_1(1-p_2)$
First tails, second heads	$(1-p_1)p_2$
Both tails	$(1-p_1)(1-p_2)$

In this simple example I prefer setting out the outcomes in the text, but for more complicated information a table has the advantage.

Classifications may be based on more than two criteria. For example, data may be collected on presence or absence of side effects (first categorization) after the administration of one of three drugs (second categorization) to either male or female patients (third categorization). When there are only two classifications, one is allotted to rows and one to columns to form a *two-way* table like Table 6.2. How does one cope with a third classification? The usual way is to divide the table (often referred to as a *three-way* table) into sub-tables; there is a *two-way* sub-table for each level of, or grouping, forming the third categorization. For the example on the side-effects of three drugs one way is to form a two-way sub-table showing the drug and side-effect categorizations for males and another two-way sub-table showing these for females. This is done in Table 6.5, where the sub-tables are presented side-by-side.

Table 6.5 Incidence of side-effects among male and female patients, each treated with one of three drugs

	Males			*Females*	
	No side effect	Side effect		No side effect	Side effect
Drug A	27	2	Drug A	35	7
Drug B	18	0	Drug B	22	4
Drug C	11	9	Drug C	19	0

This table shows clearly that the incidence rates of side-effects differ between sexes and that these differences depend upon which drug is administered.

Tables are often more extensive than those given here, but avoid, if possible, tables extending to more than one page. If a planned table spreads over two or more pages try to rearrange it into smaller tables. For example, if you want to display data showing the population of all major cities in the world broken into six different age groups, you might give separate tables for each continent. Keep detail in a table to what is needed, but see that it contains as much useful information as it should. For example, in Table 6.5 because differing numbers of patients have been treated with each drug, it might be useful to append a column in each sub-table giving total numbers of males and of females treated with each drug. One could also add an extra row giving total numbers for each sex who did, or did not, exhibit a side-effect irrespective of the drug they received; however, it is hard to see what sensible interpretation could be put on these latter totals. My inclination is always to omit information that may only confuse. A reader who wanted the latter totals could calculate them from the table, even without recourse to a pocket calculator. Other arrangements of the data in Table 6.5 are possible; that in Table 6.6 is one. Here there are three sub-tables, each of two rows and two columns.

Table 6.6 Incidence of side-effects (N=none; S=side-effect) among male and female patients, each treated with one of three drugs

	Drug A			*Drug B*			*Drug C*	
	Male	Female		Male	Female		Male	Female
N	27	35	N	18	22	N	11	19
S	2	7	S	0	4	S	9	0

Tables 6.5 and 6.6 both contain the same information; which is appropriate depends upon what is required. It is easy to see from Table 6.5 how side-effect incidence rates vary between drugs within each sex because the relevant information for males is given in one sub-table and that for females in the other. Table 6.6 shows more clearly how incidence varies between sexes for each named drug because the relevant information for any one drug is contained within one sub-table.

To give a tidy layout in Table 6.6 I used abbreviations, N, S to indicate presence or absence of a side-effect and explained this in the heading. Unless you are preparing CRC, a decision about whether to use abbreviations may be left to the copy-editor or whoever marks up the copy for the typesetter.

Do not cram so much into a table that it looks unduly crowded. A common fault is to give data to an unnecessarily high (and often spurious)

level of accuracy. To claim that the population of a certain city on 1 Jan 1993 was 2 412 317, whether in a table or in the text, implies spurious accuracy. Precise definition of a city *population* is difficult. Are visitors or temporary residents included? What constitutes a temporary resident? Are those temporarily absent to be counted? There is also the impracticality of making an exact physical count on a given date. On 1 January, if the city were in a country that celebrates New Year seriously, some census takers might have too massive a hangover to count accurately.

If we are told that the number of citizens receiving retirement pensions at that date is 317 204 this may be correct, because the authority paying these pensions should have accurate computer-based records of who receives benefits; however, this figure is unlikely to be the number entitled to these payments, for there are likely to be some who are entitled but do not claim and perhaps also fraudulent claimants.

Table 6.7 illustrates a situation where the data may be accurate in the sense just described at some date in each given period, but the high precision makes it hard to see the wood for the trees.

Table 6.7 Numbers receiving retirement and invalid pensions in Australia

Year	Retirement	Invalid
1974–5	1 097 225	168 784
1976–7	1 205 347	202 963
1978–9	1 292 476	219 843

Source: Australian Bureau of Statistics *Pocket Year Book*, 1981

Clearly there has been an increase in numbers of claimants for both categories of pensioner between 1974–5 and later years. What may also be of interest is the *relative* or *percentage* increases. For example, is the rate of increase greater for retirement or for invalid pensioners? It would be easier to see this if we rounded the numbers of recipients to the nearest thousand. At the same time, so that people can see both the increase and the rate of increase, why not also give for each later period the percentage increase relative to 1974–5 if these are relevant? I do this in Table 6.8 which shows clearly that although the numbers of invalid pensioners are appreciably less than the numbers of retirement pensioners, numbers of the former are increasing at a greater rate. In passing, note that I calculated percentage increases using the data in Table

6.7. If the rounded counts (to nearest thousand) in Table 6.8 were used, there would be slight, but not important, differences in the first decimal place.

Table 6.8 Numbers of recipients (in thousands) of retirement and invalid pensions in Australia and percentage increase in numbers compared to the 1974–5 base

Year	Number of recipients		Percent increase	
	Retirement	Invalid	Retirement	Invalid
1974–5	1097	169		
1976–7	1205	203	9.9	20.3
1978–9	1292	220	17.8	30.1

A footnote in Table 6.7 gives the source of the data; give this when reproducing part, or all, of published tables. It is usually necessary also to obtain permission from the copyright holder. However, the official statistical services in many countries allow 'brief' quotations of official statistics providing the source is acknowledged.

Sound advice on presenting tables is given by Ehrenberg (FRIII).

6.3 GRAPHS AND DIAGRAMS

Books have been written about graphs. I mentioned in section 2.2.4 the excellent account in Edward Tufte's *The Visual Display of Quantitative Information.* (FRIII). I give only a broad survey because the need for, and the best way, to use these tools depends upon the subject matter and the type of book you are writing. Some applications, particularly in engineering, are highly specialized, involving, for example, diagrams showing details of complex circuits. In section 2.2.4 I indicated a few ways to use graphs and diagrams. There are many other possibilities; well-known types of graph include pie-charts, histograms, bar charts, scatter diagrams and other data graphs; some are available in two- or three-dimensional form. Sometimes there is no clear distinction between the terms *graph* and *diagram*, e.g. decision trees, networks or flow charts are described either as graphs or as diagrams. Descriptive pathways, like those in Figures 6.3 and 6.4, are best referred to as diagrams. Drawings that illustrate for example, circuit layouts, pieces of equipment, or the structure of an organism, are purer forms of diagram.

Maps are another familiar type of diagram. These range from those produced by the UK Ordnance Survey and equivalent organizations in other countries, to city street plans and the familiar weather map, as well as geological and other charts showing special features.

In books, manuals, research papers or reports, both graphs and diagrams are usually referred to as *figures*. When numbering and referring to these consistently use either *Figure* spelt out in full, or the abbreviation *Fig.* Do not use one in captions and the other in the text; follow house-style if a preference is stated.

Depending upon your subject and your interests, you will have met some or all of the graph and diagram formats I have mentioned, and perhaps others. Icons are increasingly popular and are often used as language-independent signs to convey information without words, or to supplement or emphasize a word description. Well-known examples include 'international' traffic signs, the familiar no-smoking sign, and the male and female silhouettes designed to ensure you choose the right door even if you do not know your *signore* from your *signora*. Icons — e.g. a lightning flash to indicate a warning or need for caution — are used in many instruction manuals. Many graphic and word-processing software packages include commonly-used icons in 'clip-art' files. Figure 6.1 shows familiar examples.

Computer graphics packages simplify the production of appropriate graphs or general diagrams. Consult your publisher, or refer to a style guide, to find the required form. High quality (e.g. laser) prints are essential for photographic reproduction. Some spreadsheet or word-processing packages, given specific data, produce a variety of graphs automatically. Unfortunately, in such graphs the axes are often not well positioned or are not clearly labelled, and lines may not be of the required thickness; adapting the diagrams to a form suitable for publication may take almost as long as preparing a fresh graph.

For labels on vertical (y) axes present these, if possible, in the 'landscape' format illustrated in Figure 6.2(a). Avoid the 'vertical pile' format of Figure 6.2(b); it is hard to read and not pleasing to the eye. Figure 6.2(c) represents a compromise that is less acceptable than the form in Figure 6.2(a).

The captions to Figures 6.3 and 6.4 give the source of the figure. As for tables, you must do this if you base a figure on somebody else's work. If you reproduce a figure from another source copyright permission and due acknowledgement (Appendix A) is also required.

Figure 6.1   Some familiar icons used in technical and business writing.

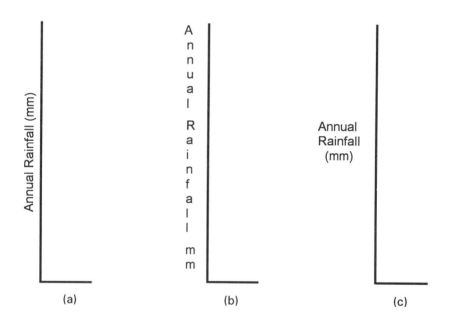

Figure 6.2 Vertical axes are best labelled as in (a) in preference to the ways shown in (b) or (c).

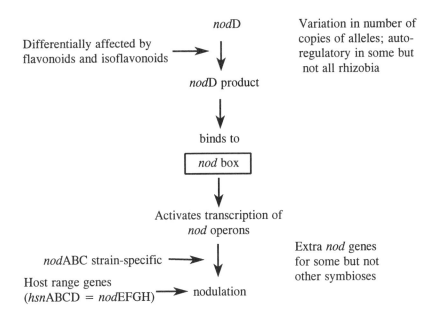

Figure 6.3 Outline of ways in which *nod* genes in rhizobia may act, from Sprent and Sprent (1990).

In Figure 6.3 the text is in the same font (*Times Roman*) as the body of the text in this book, whereas in Figure 6.4 it is in a contrasting sans serif font (*Arial*). Some authors (and publishers) like diagram text to match that in the body of the book — others prefer a contrast. Whichever you choose, be consistent; the different fonts here are purely for illustrative purposes.

Graphs or tables are often valid alternatives (section 2.2.4). Figure 6.5 is a bar chart representation of the data in Table 6.5 The format of Table 6.5, where there were two sub-tables, is retained by using separate bar-charts corresponding to each sub-table. The caption is sufficiently detailed to make the figure interpretable without referring to the text. Avoid vague captions like:

Figure 6.5 Drugs and side-effects.

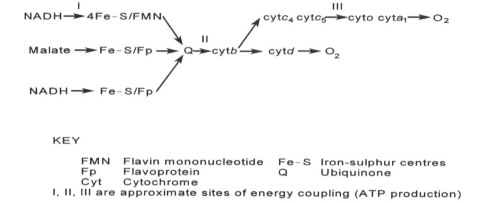

KEY

FMN	Flavin mononucleotide	Fe-S	Iron-sulphur centres
Fp	Flavoprotein	Q	Ubiquinone
Cyt	Cytochrome		

I, II, III are approximate sites of energy coupling (ATP production)

Figure 6.4 Electron transport pathways in *Azobacter*. From Sprent and Sprent (1990), after Haddock and Jones (1977).

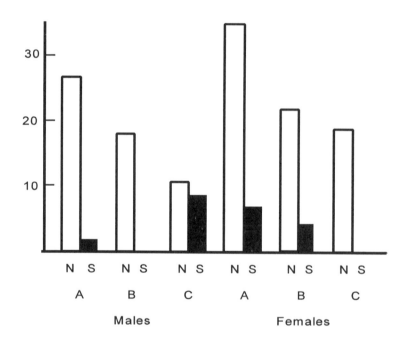

Figure 6.5 Bar chart showing numbers exhibiting absence (N) or presence (S) of side-effects for each of three drugs (A, B, C) for male and for female recipients.

Flow charts are widely used to illustrate logical processes such as computer algorithms. Different box shapes are used conventionally to indicate various logical states — e.g. oval boxes for start and finish, rectangular boxes for actions and diamond-shaped boxes for decision points. There are no routes into, and only one route out from, a *start* box and one or more routes into, but no routes out from, a *finish* box. *Action* boxes have one or more routes in, but only one route out (after action is taken). A *decision* box has one or more routes in and always two routes out. Figure 6.6 is a flow chart for an algorithm that reads, in order, a set of ten numbers into a storage location denoted by X, i.e. the current content of X is the last number read in. After all ten numbers are read, the mean is printed and the program stops. If you are not familiar with computer programming and terminology you may prefer to study the alternative flow chart, Figure 6.7, representing the procedure I suggested in Chapter 4 for seeking a publisher.

Diagrams that superficially resemble flow charts are used in operational research techniques such as critical path analysis, where the aim is to schedule a sequence of jobs to complete a complex task in the minimum possible time. I do not give details, but this and related networking problems such as determining a best route to minimize transport costs, are specialized applications and the principles are familiar to project engineers and operational research experts.

Figure 6.8 shows a way of representing periodic data. The basic data are the total numbers of deaths in one month over each hourly interval in a day recorded for a large hospital. Each closed circle (●) plotted for the relevant hourly interval outside the larger open '24-hour clock' circle represents a death. The top of the clock represents midnight and proceeding clockwise, each quadrant corresponds to a six-hour period. Not unexpectedly, there is a high death rate in the late night and early morning with an irregular scattering of deaths in other hourly intervals. This is one of several useful techniques for plotting periodic or directional data of a kind that arise commonly in biology (e.g. the directions of animal movements) and geology (e.g. orientation of rock faults), and many examples using real data are given by Fisher (1993).

Figure 6.9, our final illustrative example in this section, is a familiar pictorial graph using an icon (here a telephone) to indicate the number of new connections to a network (each telephone representing 100 connections) in several years. In this diagram the information is effectively given to the nearest 100 for each year. If a more precise figure is required (say to indicate 450 connections) this can be achieved by showing four complete telephones and a final 'half' telephone. Clearly, because of the shape of the telephone,

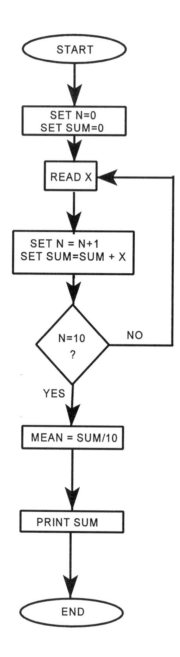

Figure 6.6 Flow chart for a program to read ten numbers and to calculate and print their mean.

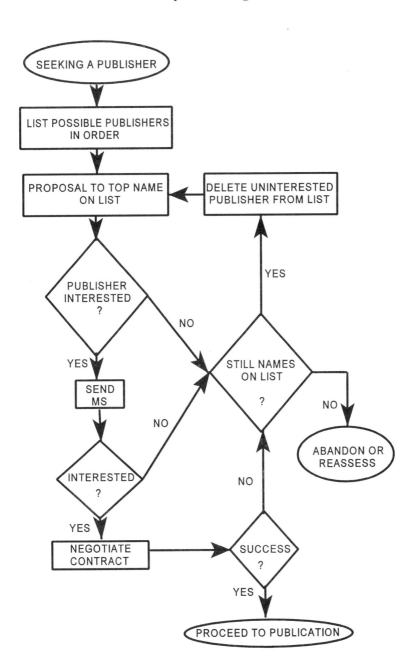

Figure 6.7 Flow diagram for the procedure for seeking a publisher suggested in section 4.1.1.

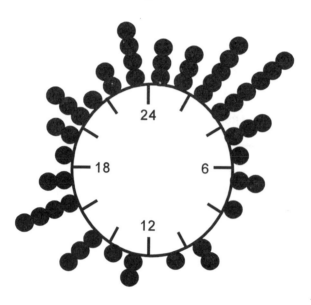

Figure 6.8 A circular representation of numbers of deaths in each hourly period of a day totalled over one month at a hospital. Each circular symbol outside the clock represents one death. The large clock circle represents a 24 hour clock.

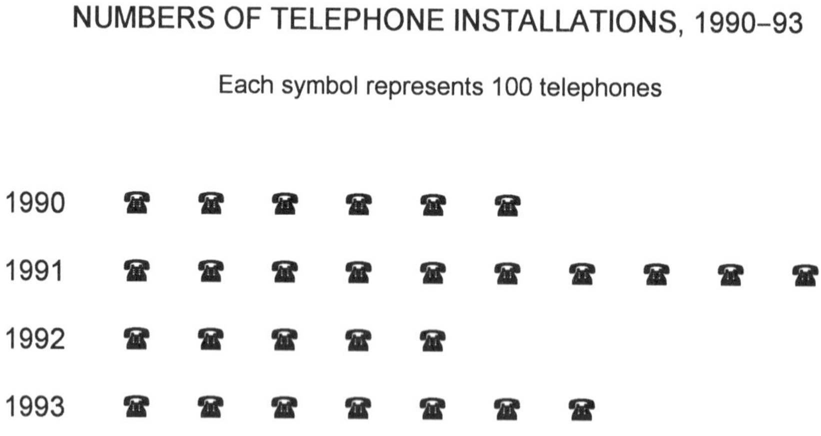

NUMBERS OF TELEPHONE INSTALLATIONS, 1990–93

Each symbol represents 100 telephones

1990	☎	☎	☎	☎	☎	☎				
1991	☎	☎	☎	☎	☎	☎	☎	☎	☎	☎
1992	☎	☎	☎	☎	☎					
1993	☎	☎	☎	☎	☎	☎	☎			

Figure 6.9 A symbolic graph using an icon of a telephone to indicate numbers of new telephones connected to an exchange in each of four years.

connections of less than 100 units, say 73, can only be indicated approximately by showing about three-quarters of a telephone. This type of graph is often useful in school textbooks and also in books intended for the general reader, or in technical articles aimed at a wide audience.

What are appropriate graphs and diagrams depends on the type of book and the subject matter, but the following reminders are relevant in many contexts.

- In data graphs like Figures 2.1 and 2.2 label axes clearly; break axes if the origin is not at the intersection. Choose scales so that the main features of your graph fill the diagram area.
- In histograms, bar charts, pie-charts etc. use sufficient labelling to make clear what all sectors of the graph represent. When appropriate, show numerical scales clearly.
- For all graphs and diagrams provide a descriptive caption to make the graph intelligible without reference to the text (although you will normally discuss particular aspects in the text).
- Keep diagrams simple, but provide all information relevant to their interpretation. For specific types of chart stick to established conventions; e.g. those given above for a flow chart.
- Shading is a useful device, but with computer generated diagrams, shading that looks satisfactory on a laser print does not always reproduce well in a photographic printing process. This problem may be overcome by omitting such shading and replacing it by a screening process at the printing stage. Seek the advice of your publisher before using shading.
- General diagrams include sketches of laboratory equipment, machinery, etc., and also maps, plans and circuit diagrams; maps may be especially useful in meteorology or geology.

6.4 PHOTOGRAPHS

Seek your publisher's agreement at an early stage if you want to use colour photographs, because these may increase book production costs dramatically (section 2.2.4). This also applies to black and white photographs if very high resolution is needed to show specific features such as deformation of the cell walls of an organism or the effects of corrosion on a metal surface, or important features that can only be shown by techniques such as electron microscopy. For colour photographs high quality paper is used and the photographs are often grouped on consecutive pages and each item is called a *plate*. These are numbered Plate I, Plate II, Plate III, . . ., in a sequence separate from that used for figures. If photographs are used, find out from your publisher in advance the technical requirements both for colour and black and white. I gave some common specifications in section 2.2.4.

6.5 MATHEMATICAL AND CHEMICAL FORMULAE AND EQUATIONS

Except for short symbolic expressions, e.g. $E=mc^2$, that may be included, as here, in an ordinary line of text, formulae and mathematical or chemical equations are displayed. The resulting break in the text is an incidental device for improving layout. However, the complexity of formulae and equations arising from special symbols, brackets, subscripts and superscripts, carries dangers of inaccuracy or ambiguity. This is a common area for 'misprints'. Vital brackets or parentheses (the latter are technically round brackets, but mathematicians often refer to them all as brackets) are easily omitted, or mistakes are made with signs or symbols, e.g. plus and minus, equality, inequality, an *a* instead of an α. These errors are particularly serious in textbooks because many students have an unjustified faith in the printed word.

Special computer software is available for preparing scientific or mathematical text; one such package with outstanding flexibility is *TeX*, described in some detail by Clark (FRIII). Some word-processing packages, e.g. *WordPerfect 5.1* and later versions, include facilities for setting mathematical formulae. Successful use may depend upon the capabilities of your printer. If your word-processor or printer cannot deal with mathematical, or other symbolic, formulae, it is best to insert these in a manuscript by hand, putting notes in the margin to guide the printer with unusual symbols, e.g. Greek letters that look like their Roman counterparts, or special mathematical symbols like ∇ or $\forall$. If you do this, check carefully at the proof-reading stage that each has been set correctly.

Several word-processing packages let you set expressions like:

$$P=\{\Sigma_s(a_sc_s)\}/(2c_+^2) \qquad Q=\{\Sigma_s(a_sd_s+b_sc_s)\}/(2c_+d_+)$$

or

$$\theta*= \frac{\Sigma_s n_{11s}\, n_{22s}\, /N_s}{\Sigma_s\, n_{12s}\, n_{21s}\, /N_s}$$

providing your printer can reproduce them!

Publishers of books on mathematics, statistics, physics or engineering often give guidance on how to present mathematics. Difficulties may occur even with common fractions in text; e.g. is one-half better represented by 1/2 or by ½?

Chemical equations, like mathematical formulae, are usually displayed. If the equation is to indicate molecular structure, it is best to use a diagram and this may be generated with graphical software. Figure 6.10 gives an example, again taken from Sprent and Sprent (1990). Because it is not an equation, but indicates a synthesis, an arrow is used in place of an equality.

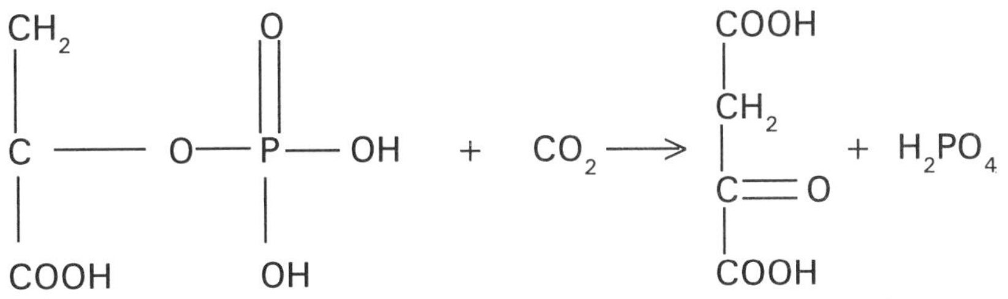

$$
\begin{array}{c}
CH_2 \\
|
\end{array}
\quad
\begin{array}{c}
O \\
\|
\end{array}
\qquad\qquad
\begin{array}{c}
COOH \\
|
\end{array}
$$

$$
C \;\text{—}\; O\text{—}P\text{—}OH \;+\; CO_2 \;\longrightarrow\; \begin{array}{c}CH_2\\|\\C\!=\!O\\|\end{array} \;+\; H_2PO_4
$$

$$
\begin{array}{c}|\\COOH\end{array}\qquad\begin{array}{c}|\\OH\end{array}\qquad\qquad\qquad COOH
$$

Fig. 6.10 Synthesis of oxaloacetic acid by the PEP carboxylase reaction.

6.6 EXAMPLES AND EXERCISES

Textbooks in subjects like mathematics, computer science, physics, engineering and business studies, often need worked examples and student exercises. A good selection may tip the balance in favour of a book that, in other respects, is well matched by competitors. Essential features are

- variety;
- careful grading.

Real life examples are ideal, but reality is usually too complicated to illustrate one particular point at a time. In these circumstances make the context realistic, even if it has to be simplified.

Old fashioned arithmetic books used to delight in problems like:

The hot tap alone fills a bath in 6 minutes and the cold tap alone fills it in 4 minutes. The waste pipe will empty the bath in 5 minutes if neither tap is turned on. If I turn on both taps and do not insert the waste plug, how long does the bath take to fill?

This example tests the understanding of an important principle, but it is unrealistic, and as such, often irritates pupils. What idiot would turn on the taps without inserting the plug?

Here is the same principle tested in a realistic environment:

> A reservoir is supplied by two inlet pipes, A and B. There is one outlet pipe, X.
> If X is closed and only A is open it takes 6 days to fill the reservoir and if only
> B is open it takes 4 days to fill it. If A and B are both closed, but X is open it
> takes 5 days to empty the reservoir. If the reservoir is initially empty and A, B,
> X are all open, how many days will it take to fill the reservoir?

This is realistic because most reservoirs have one or more in-flows and out-flows; some, or all, may function at the same time. At a more advanced level realism could be increased by allowing the rates of inflow and outflow to vary from day to day. A useful ploy, particularly with exercises for students, is to build up a series of related problems, each introducing additional features. For instance, the reservoir example above might be followed by a further exercise in which the rates of inflow and outflow change when the reservoir became half full. Such a change might be realistic in practice because an increased rainfall accelerated inflow, or if, when the reservoir became half full, the authorities removed certain restrictions such as a ban on the use of garden hoses and this increased outflow.

Avoid exercises that repeatedly invoke the same idea where, if students know how to do one, they can do the lot. All too often one sees a string of exercises differing only in environmental or numerical detail. These four exercises all test the same elementary notion about probability:

1. There are 15 boys and 21 girls in my class. If one of them is selected at random to present a bouquet to the President's wife, what is the probability a boy will be chosen?
2. A bag contains 16 blue marbles and 25 red marbles. I put my hand in the bag and select one marble. What is the probability it is blue?
3. There are 6 green cars and 4 white cars in a motor show-room. If one of these cars is chosen at random, what is the probability it is green?
4. My mother has 7 blue jumpers and 5 pink jumpers. She is very untidy and each morning she grabs the nearest one. What is the probability she grabs a pink one on Tuesday morning?

There should be only one, or at most two, of those exercises, even in a very elementary statistics text book.

Avoid boring examples or exercises even if they are built around real data. I once based a series of examples and exercises on counts of random samples of books on my bookshelves. The data were real, but of little interest even to me! Had they been data for samples of books from the British Library on a range of topics, readers might have been more interested.

6.7 LABORATORY EXPERIMENTS

If you write a textbook, perhaps even one for background reading or for the interested amateur in a subject like physics, chemistry or biology — or in geology or practical aspects of engineering — you may include descriptions of laboratory, or sometimes field, experiments. Select your experiments carefully. They should have relevance to the basic theme of your text, be within the competence of your target readership and be such that many readers are likely to have the facilities to carry them out. Make descriptions clear and accurate. Incorrect or incomplete instructions that endanger students or others following your directions may land you and your publisher at the sharp end of an action for damages; a situation more beneficial to lawyers than to the advancement of science. Do not forget, for example, that there is a relatively safe and also a highly dangerous way of diluting concentrated sulphuric acid; that certain precautions must be taken (and are legally required) when high-voltage electricity is involved. If there is a potential danger, stress that any experiments described should only be performed under the supervision of qualified laboratory staff, perhaps adding a disclaimer for shortcomings or inaccuracies in your description of experiments or techniques, although this will not exonerate you or your publisher from liability for the consequences of a careless or blatantly misleading statement. How seriously publishers now view the possibility of legal action is indicated by this clause in my contract for this book:

> The Author shall include in the text of the Work appropriate warnings concerning any particular hazards that may be involved in carrying out experiments or procedures described in the Work or involved in instructions, materials, or formulas in the Work, and shall explicitly mention relevant safety precautions, and give, if any accepted code of practice is followed, a reference to the relevant standard or code.

So, to protect myself, I warn you now that even if you follow all my advice, there is no guarantee that you will find a publisher for your work.

6.8 RESEARCH AND COLLATING INFORMATION

In this chapter I have described features often peculiar to, or of special relevance to, various types of scientific and technical writing. To use these effectively, as well as for other facets of your writing, you will often have to seek or recall information. Much of what I say in this section is particularly relevant for obtaining material to illustrate specific points, or for finding data for examples or exercises.

What background research an author has to do depends both on the type of book and on the author's existing store of knowledge. A *want-to-do* writer is unlikely to start writing a book without a reasonable amount of relevant knowledge at his or her fingertips. However, even experts writing research monographs are unlikely to have all the material needed immediately to hand. If there is a set of data to support a certain theory you may know its general nature and that it is in a paper by Smith that appeared in the *Proceedings of the Royal Society* in either 1989 or 1990; if the paper is to be quoted, or data from it are to be used, a precise reference is needed. If you are methodical you may have a card-index system or a computer database file that will give you the relevant information immediately. Otherwise you face a library search of the likely volumes of those *Proceedings* to find the required paper. The position is more complicated if all you can remember is that the information you want was in a paper by one of several possible authors and that it was published in one of five journals some six to ten years ago.

Even experts writing any type of book in their own field often have a hard job selecting appropriate material. In the next chapter I discuss a possible layout for a textbook that includes a chapter on field aspects of nitrogen-fixation by plants, suggesting that this might be illustrated by case studies to show the diversity of the process. There are literally hundreds of relevant case studies in the botanical literature. You may only want a dozen examples. It may take a long time to select a dozen cases that show the required diversity. You will bore your readers if all your examples involve beans grown in the American mid-west, or sub-tropical clover pastures in Australia.

To save repeating work, take notes of any point that might be useful — be it snippets of information, an interesting reference, a new theory, a novel approach to a problem — as soon as it comes to your attention, even if it is not of immediate relevance. It is frustrating, if on reading a paper to get certain information, you see another point you feel may be worthy of mention in a later chapter, only to find when you settle down to write that later chapter that you cannot remember in which of half a dozen papers the point you wanted to make was raised. This is when lazy authors are tempted to use that platitude *It is well-known that . . ..*

It is easy to ramble on about how to collect and store information. Writers each have their own approach. Build a database that suits you. In doing so you will find it essential to refer to other databases, and retrieving any relevant data in the form that suits you best will make your task of writing easier. In many subject areas not only are there the classic conventional databases such as libraries and archives, but a growing number of electronic data bases, some on disc, some on CD-ROM and others accessible on computer networks.

In section 6.7 I emphasized the importance of realistic examples and exercises where appropriate. Published work is often a good source of suitable material. If you quote extensively from published work you will need copyright clearance (Appendix A), but if you base a student exercise on small data subsets from published work, an acknowledgement of source will usually suffice. Seek your publisher's advice if in doubt and also read Appendix A on copyright.

Many kinds of books

7.1 CATEGORIES OF BOOKS

I pointed out in section 1.2 that writing monographs, textbooks at various levels, background reading and popular books each call for different writing styles. They are not hard and fast categories. Roger Penrose's *The Emperor's New Mind* was aimed at the broad market consisiting of those interested in, but with no specialist knowledge of, artificial intelligence. It is also sound background reading for students studying that subject. Darrell Huff's classic *How to Lie With Statistics* probably had, and certainly achieved, the two objectives of entertaining the layperson and educating the suppliers, users, abusers and students of statistics. A textbook on basic calculus written primarily for engineers might, if it took a suitable approach, also be useful background reading for mathematics students without a specific interest in engineering, or for a layperson keen to learn the rudiments of calculus.

This chapter is about some features specific to various types of book.

7.2 MONOGRAPHS

The name *monograph* was originally used for a treatise dealing with a particular species or group of plants or animals, but it now covers any authoritative dissertation on a clearly defined and specific branch of learning, or an account of theories or techniques relevant to a particular type of study. The author is usually an expert in the subject area.

The target readership is generally people with a serious interest — and probably some expertise — in the field, or it may be people working in cognate areas who need or seek a concise account of some topic or method.

A research monograph usually collates the latest findings in the subject area when these have hitherto only been available in scattered research reports. As well as collating this material, critical evaluation of the results may be a valuable component.

There are other kinds of monograph. There may be a well-established corpus of knowledge about some topic which, although only being advanced slowly by current research, suddenly becomes important because of new applications. A monograph that links the established material to the new applications may then be appropriate. Also, if a topic has been developing slowly and in a fragmented way, a monograph to summarize and relate these developments often proves useful.

Monographs are sometimes written to describe new laboratory techniques for detecting, say, oral cancer; or to describe new theoretical methods for tackling a particular problem, e.g. statistical techniques for estimating moorland grouse populations, or perhaps bird populations in general. There may be a fine line between a monograph of this type and a sophisticated instruction manual; the latter is likely to put emphasis on how to apply techniques or theories, while the former will almost certainly devote more attention to the rationale behind them.

Preparing a monograph is, from a 'writing technique' viewpoint, one of the less difficult forms of scientific writing; but I am not equating *less difficult* to *easy*. Usually the target readership is clearly defined — those who want, or need, to know about the subject matter — and one may often assume readers will have specific background knowledge. For example, if you are writing a monograph about plants exclusive to Wales, unless you are writing for the 'popular' end of the market spectrum, most of your potential readers are likely to be familiar with general botanical terms; or if you are writing a treatise on the special theory of relativity, most of your readers could be expected to have a good background in theoretical physics. In cases like that you need not explain jargon or special concepts that will already be familiar to target readers. There will be borderline situations where the best solution may be to define certain terms in a glossary to help any reader unfamiliar with a concept or terminology. You will need to explain ideas or jargon peculiar to the restricted subject matter of your monograph if you are not able to assume your target readers are already familiar with the field.

Common faults are obscurity and failure to explain in sufficient detail concepts which, although familiar to you, will not necessarily be known to all readers. Be particularly careful if your monograph takes a 'new' approach to an established discipline. For example, if you are writing about biochemical aspects of cell division, aiming to explain its importance to biologists with little knowledge of biochemistry, you should either explain the relevant biochemistry, or give references to sources where that material is treated at an appropriate level. Quote sources that are likely to be readily available to readers. It is not helpful to recommend a book that has been out of print for ten years, unless you are certain it is still freely available in libraries. Remember, too, that what is available in libraries in your country may not be so readily obtainable in other countries.

A terse, but not too abrupt, style is often appropriate for monographs. Your readers are looking for information; some of them may only be interested in specific aspects of the topic you are dealing with; take this into account in ordering your material. For example, if you write about plants exclusive to Wales and expect some of your readers to be interested only in plants found in Snowdonia, you might devote a chapter to that area — but do not do this if it would conflict too violently with the layout of the rest of your book. A good index is a great help to 'selective' readers. For this, or any writing, it is a good idea to get a potential target reader to comment upon your draft manuscript.

If your monograph deals with a theory or a new experimental technique that might be applied in several disciplines, e.g. in biology, psychology, engineering and quality control, consider starting the introductory chapter with simple examples, which need not be discussed in great detail, from each of these fields.

7.3 STUDENT TEXTS AND BACKGROUND READING

I cover books both for postgraduate and for undergraduate students in this section, leaving school textbooks to section 7.4. At the postgraduate level there may, in a specialized subject area, be little difference between a monograph and a formal postgraduate text, or between the latter and a book for background reading. Indeed, many monographs are required reading for postgraduate students; also postgraduate texts that cover broad aspects of a subject, e.g. a postgraduate text on biochemistry, or on design principles in civil engineering, may also be background reading for other students in these subjects, even if they do not use the book as a formal text for a lecture course or study program. Again, there is often no clear divide between texts used by postgraduate students and those used at advanced levels (e.g. the final year) of undergraduate courses. It is because of these overlaps that I decided to take this general look at preparing textbooks and background reading at the tertiary education level, pointing out differences in approach appropriate at the various stages.

7.3.1 Motivation

For textbooks and background reading in widely studied subjects — e.g. chemistry, biology, physics, mathematics, psychology, engineering, economics, business studies — the market is competitive. Would-be authors are often not impressed by available books and believe they can produce a better one and that the venture will have good commercial prospects. Experts — and sometimes the not-so-expert — easily convince themselves that it would be easy to write a book as good as, or better than, an established and successful one. If you feel this way, reflect for a moment. Most successful textbooks are well written and meet a precise market need. You may deplore the emphasis in a particular book — dislike its 'cook-book' approach, or think that it highlights the wrong points — but if that book is widely used and popular with teachers and students, then a novice author — even an experienced writer — will be hard pressed to produce an equally marketable text, let alone one that will sell better.

One problem is that a new book which is only as good as a market favourite has to battle built-in inertia. Lecturers or teachers will be reluctant to change, recognizing the repercussions in the second-hand market for the established text — a market that may not bring joy to authors or to publishers — but one looked upon sympathetically by teachers because of its importance to impecunious students.

Even if your book is superior to an established text, it may have to be considerably better — or much cheaper — if it is to be a serious competitor.

Successful new texts include those that plug gaps in coverage by current favourites, providing that they also deal competently with topics that are already well handled. Major syllabus changes also stimulate a demand for new books, as do technological developments or new attitudes towards the way in which things should be taught. In these situations existing texts may suddenly become obsolete. For example, in the late 1980s and early 1990s some biology textbooks dated very quickly as a result of developments in molecular biology and genetic engineering. Again, modern computing facilities have revolutionized the teaching of, even elementary, mathematics; cheap pocket calculators sounded the death knell for approaches using tables of logarithms and trigonometric functions. In rapidly developing subjects like computer science and information technology, changes in what is important are so frequent that books on these topics are in danger of becoming obsolete even before publication. In subjects like architecture and civil engineering, less dramatic but frequently occurring changes like the introduction of new building materials and novel construction methods, even changes in building regulations, must be taken into account in planning courses and often provide scope for innovative texts.

Fashionable courses like those that proliferated on both sides of the Atlantic and elsewhere in the late 1980s and early 1990s in business administration — the MBA — created a demand for specially slanted texts covering a range of topics from accounting to operational research and personnel management; but if you have aspirations in that direction, beware the dangers of jumping on overcrowded bandwagons.

Given the right niche in the market, producing a good text, especially at the undergraduate level, may be academically, and sometimes commercially, rewarding to a lecturer or teacher with enthusiasm for a subject. Finding that niche depends upon:

- your discipline;
- the competition;
- factors like changes in technology or attitudes towards how a subject should be taught.

Be innovative, providing you have confidence in what you are doing. New laboratory equipment, developments both in computer software and hardware, new audio-visual aids in the classroom, may all suggest better ways to teach your subject. Do you see a need for a book that includes relevant computer program listings, or for one that is sold with a disc containing programs ready for students to use?

7.3.2 Some fundamentals

Always remember the basics of *macrostyle* and *microstyle* outlined in chapters 2 and 3. First macrostyle. Have you defined your target readership? Thought about what jargon you can assume your readers will know, and how much you will have to explain? Decided upon how, and in what order, new concepts should be introduced? The best logical order for the main topics?

Other aspects of *macrostyle* calling for early decisions are whether to include student exercises; how freely to use illustrative examples; whether to include tables, graphs or diagrams; whether you will describe laboratory experiments; if you will divide chapters into sections and perhaps also subsections.

When you start writing, emphasis shifts from *macrostyle* to *microstyle*. You have to choose those right words and put them in the right places, at the same time keeping an eye on pace. Remember also that some, but not too much, redundancy, helps readers digest new concepts.

Typically, if you are writing an undergraduate text, you will have experience of teaching a course at the appropriate level. A good textbook acts as an anchor for many courses. One that is attractive, clear, easy to read, yet authoritative, gives a teacher freedom for innovation as well as providing useful back-up for a student who finds the subject difficult. Bear these points in mind as you write.

One useful microstyle device, particularly in introductory textbooks or those for a more general readership, is the seemingly casual insertion of information that is likely to be known to some but not all readers. This may sometimes be done by putting several related pieces of information in one sentence, or by slipping in an explanatory clause at an appropriate point. For example, if you are writing about the Mississippi river, the sentence

The Mississippi river flows into the Gulf of Mexico southeast of New Orleans, 3779 km from its source in Lake Itasca, Minnesota

gives three pieces of basic information about that river. Some, but not all, of your readers may know them all. Linking this related information into one sentence is less laborious than putting it in three separate ones. Slipping in information casually is also useful if you use a piece of jargon or terminology that you expect most, but not all, readers to know. You might write:

Let m *be the slope of the line, i.e. the tangent of the angle between the line and the* x-axis

providing you are happy that this explanation will suffice for most readers without the need for a formal definition of *tangent*. You might add a simple diagram to help. However, if you expect some readers to know little mathematics a more formal explanation of the terms slope and tangent may be needed. The 'slip-in' device is also useful for instant translation of a foreign word or phrase, e.g.

Euclid's 'pons asinorum', *the bridge of fools, was the fifth proposition in that famous geometer's first book of the* Elements.

Prefer this to a translation in a footnote.

7.3.3 The planning stage

Here are two examples of the way a new textbook might evolve; the first in the field of biology, the second in economics and business studies. Try to think of parallels in your own discipline. If you already have an idea for a book, consider how you might adapt these approaches.

In the first example, I assume you, the author, to be a biologist who gives a course at second-year undergraduate level on the role of nitrogen in plant growth. Several existing books cover aspects of this topic. This may surprise the non-biologist, but it is because nitrogen has a key role in plant growth; also the subject is complex because plants cannot directly assimilate nitrogen from the atmosphere.

You want to write because existing books, good as they might be, are not matched to your course. Some are too advanced, others specialize in particular aspects of the subject such as biological fixation of nitrogen, or upon biochemical aspects of how plants use nitrogen. If similar courses are taught in other institutions, bear this in mind as a potential extended market; consider possible modifications that do not defeat your objective, but that might make the work attractive to these outlets. An early step is to decide upon the topics to be covered. You might jot these down in note form.

1. Natural sources of nitrogen in forms usable by plants.
2. Nitrogen fertilizers.
3. Biological fixation of nitrogen.
4. How plants use nitrogen.
5. Practical implications for agriculture, forestry, etc.
6. Other matters (a) measurement of up-take of nitrogen by plants, (b) evolutionary aspects, (c) role of nitrogen in soil regeneration, (d) future trends.

This is a framework for detailed planning. Items 1 to 3 cover the main sources of nitrogen available to plants. Your introductory chapter might outline the relative contributions from each; also in that chapter you might indicate some advantages or disadvantages of each, e.g. natural sources, such as electrical storms, only provide irregular supplies; fertilizers often act quickly, but may pollute streams or water-supplies; biological fixation reduces pollution problems and production may often be tailored to plant needs, but a key restriction is that only a few species are able to fix nitrogen. You might compare the economics of, and energy requirements for, making nitrogen fertilizers and for biological nitrogen fixation.

For a biology course, item 3 is a topic that would be extended appreciably. More than one chapter might be devoted to the several aspects of biological fixation. One chapter might list species able to fix nitrogen and describe briefly how the process works and the role of bacteria in converting atmospheric nitrogen to a form that can be used by plants, as well as where and how this takes place. A further chapter might discuss the biochemical aspects of fixation, while yet another might describe the role of nitrogen in growth and cropping for those plants that do their own 'fixing', and also ways in which other crops may benefit indirectly by return of some of this nitrogen to the soil.

Case histories might be used to illustrate practical implications for agriculture and forestry, but choose your examples carefully; readers will be bored if too many similar ones are included. A range of cases is available covering agriculture, horticulture and forestry in both temperate and tropical regions. For example, nitrogen-fixing trees grown on low fertility tropical soils provide essential fuel for cooking and heating in parts of Africa. Sensible use of crop rotations and practices like inter-planting different species, may enrich the nitrogen content of soils in areas where economic or social factors (e.g. cost or danger of pollution) preclude widespread use of artificial fertilizer.

Careful judgement is needed in developing miscellaneous topics like those in item 6 in your list. All may appeal to you as an author, but restraint is needed if the book is getting too long, or if detail is irrelevant for the courses you have in mind. Will saying too much about evolutionary processes bore students? Is the task of measuring plant uptake of nitrogen too advanced a topic for undergraduates? The role of nitrogen-fixing plants in regenerating waste soils such as mine slag-heaps or land devastated by floods or volcanic eruptions is of practical importance and therefore should be included if the course stresses applications.

Your next step is to plan chapter layout, summarizing in a few lines the main topics to be covered in each. Few authors adhere rigorously to their first planned layout, but having one helps sort problems about logical ordering that inevitably arise if you start writing without a plan.

Be clear about level of presentation. The book is aimed at second-year undergraduates, so you may make free use of standard biological terms that students will have met in their first-year or in parallel courses. Be careful with terminology for concepts specific to nitrogen fixation, or in describing the biochemistry of the process; here a glossary may prove useful.

Start to write draft chapters. If relevant and appropriate, use tables, diagrams, and perhaps photographs. Use these primarily to help understanding, but do not forget their secondary role in aiding layout by breaking text. Tables might list species that can fix nitrogen biologically, diagrams show how nitrogen-fixing bacteria enter plant roots, schematic diagrams describe biochemical processes and photographs illustrate a slag-heap at various stages of 'rehabilitation' using nitrogen-fixing, and other, plants.

If you have not already contacted a publisher, when you have an outline and several draft chapters, submit a proposal to one. If that publisher says no, try another; pausing first to consider whether your proposal needs modification, especially if the first publisher has made critical comments. However good your book, sometimes you may have to try yet another publisher, and another, and another, and . . ..

Here is an example from the social sciences; from the business and economics field. You plan a short text about the calculation and use of *index numbers*, such as the Retail Prices Index. You know the subject is discussed in many texts on economics or statistics, but you find the accounts are often too formal and that many concentrate on mathematical technicalities rather than upon the ways indices are constructed and used. Also, you are concerned because many students find the mathematics heavy going; you accept that students must perform some mathematical operations with indices, but feel that a more straightforward presentation of the mathematics is needed. You jot down the following list of topics to be covered.

1. Illustrate principles by a simple index tracing changes in, e.g., the price of milk.
2. Extend the idea to cover several commodities and demonstrate the need for weighting to allow for differing consumptions.
3. How is information obtained to construct well-known indices, e.g. the *Retail Prices Index*?
4. What does an index tell us? And fail to tell us?
5. Some technicalities. Different types of index, changing the base year, different relatives. Some mathematical techniques.

You may decide this is a good order for presenting the material, provisionally planning to devote one chapter to each numbered topic.

Index numbers give scope for using several devices that clarify and also improve layout. Tables, some simple, some complicated, will be a must. There may be graphs. There will certainly be formulae and equations that need to be displayed. Examples might be indented, and perhaps exercises given at the end of each chapter. If you are preparing CRC these exercises might be set in a smaller type; for an ordinary manuscript submission your publisher will make a decision like that. Definitions might be highlighted against a grey (or, if your publisher will sanction it, even a coloured) background so that they stand out like this:

> **Definition.** A percent index takes the value 100 at the base date.

If you use devices like highlighting definitions, indenting examples, displaying formulae or equations (with or without shading) do this consistently. Be careful to number tables, equations etc., correctly. It annoys readers to find there is an equation (5.4) and an equation (5.6), but no equation (5.5). Again, in this example, once a general plan is decided, produce a summary plus one or two draft chapters, then approach a publisher.

The best way to use features to help achieve attractive text layout depends upon the subject. Books in the mathematical sciences will almost certainly include worked examples and student exercises. The same will be true of many student texts in physics and engineering. In most of these, diagrams will also be included. Remember (section 6.3), good graphical discipline is important; label axes clearly, show scales. If the origin is not at the intersection of your axes, I remind you again, because this is so often ignored in graphs, indicate this by a break as I did in figures 2.1 and 2.2. Good graphs in a textbook not only improve appearance, they encourage students to adopt good graphical habits. Choose appropriate types of graph from the wide range of possible diagrams and charts including scatter diagrams, histograms, bar charts, pie-charts, tree diagrams, icons, etc. some of which I discussed in section 6.3. Many computer graphics packages produce striking three dimensional, and other charts, but do not be carried away by such technological wonders unless they are relevant.

For laboratory-based subjects, descriptions of experiments and laboratory techniques may be needed (section 6.7). Heed the warning there on the importance of clarity and accuracy and do not forget the possible consequences of incorrect or inadequate instructions.

7.3.4 Writing for a specific syllabus

Most academic and general technical higher education courses differ in detail from one institution to another. However, many vocational courses are based firmly on national examination syllabuses for qualifications recognized or administered by official bodies or professional associations. Qualifying examinations in accountancy and many technical diplomas are examples. There is a lucrative, but often well-catered-for, market for texts geared specifically to a relevant syllabus. You may deplore it, but pragmatism demands that these books are targeted at training candidates to pass, with minimum effort — though that often still means a hard slog — the requisite examination. Writing such texts has features in common with writing school texts, but the student readers will usually be more mature, and sometimes under great pressure, especially if studying part-time.

These texts often include, as examples or student exercises, questions from past examination papers. Such questions are under copyright, so you need permission from the relevant examining body before using them, and a fee may be payable. If you choose questions from past papers, make sure they are relevant to current syllabuses. Syllabus changes for professional examinations have become fashionable in recent years.

7.3.5 The self-teaching text

Self-teaching texts usually have a vocational, do-it-yourself, hobby, sporting or recreational emphasis. They are most likely to succeed if they are published in an established series. In the UK two well-known series are '*Teach Yourself* . . .'and '. . . *Made Simple*', but these are not the only collections, and there are also many series of books of this type in the USA and in other countries. General features of good textbook writing apply, but, in addition, remember that your reader is unlikely to have ready access to a teacher, so clarity is vital. Technical instructions are often helped by clear and simple diagrams. The leading publishers of self-teaching books usually have specific and often restrictive house-styles. Their editors can give useful hints and advice, but they expect their authors to be expert in their subject, to write clearly, and to have a flair for explaining the potentially obscure. Dozens of topics from plumbing to chess are already covered by self-teaching books, so study the market carefully to see if there is a need for any book you have in mind.

7.3.6 Books for service courses

If your subject is, say, mathematics or chemistry, and you want to write a book on mathematics for engineers, or on chemistry for biologists, remember that your readers are unlikely to share your enthusiasm for your subject. Do not deplore this; accept it as a fact of life. Include only material relevant for the likely reader, remembering also that there are time constraints for most undergraduates. Examples and exercises should highlight applications of the service topic in the reader's mainstream discipline. A book on statistics for biologists might well use simple examples from genetics to illustrate basic probability concepts, rather than resorting to drawing marbles of various colours from bags. However, if you are writing, say, for first-year undergraduates in biology, avoid illustrative examples that involve genetics that students will not meet until their third year. In the same book, principles of sampling might be illustrated by the problem of estimating the proportions of various species of trees in a forest, or of estimating the number of fish in a lake. If you are writing a book on mathematics and statistics for business executives they will be more interested in a discussion of the expected accuracy of a sample audit than in an account of how to estimate the number of fish in a lake (unless the estimation procedures are linked in some way to assessing the lake's potential for commercial fishing).

Similarly, a book on chemistry for engineers should deal with properties of chemicals particularly relevant to materials used by engineers; students appreciate topicality, so include, if appropriate, the chemistry of new materials coming into common use. The book might also deal with chemical problems associated with corrosion of structures.

7.3.7 Background reading for students

At the start of this section I pointed out an overlap between textbooks and books for background reading. However, books specifically for background reading differ from a core undergraduate text in that there is often less need to tie them to a particular course structure. The down-side is that it is harder to pinpoint a target readership, the up-side that it gives more freedom in treating a subject. Nevertheless, think about a target readership and how best to reach it, because the potential market may be small. Most students buy only essential books, so sales of background reading material are largely to academic, institutional, or other libraries, bodies with purchasing policies usually restricted by budgetary considerations. For UK authors writing background reading material, especially for subjects with a vocational slant, the export market may prove more profitable than the home market. American writers, however, have a larger potential home market.

To write a book designed as a textbook for some courses, and as background reading for others, needs something of a balancing act; a blend of judgement and fortune may be a key to success. I got it wrong in the early 1980s, when I became interested in a statistical topic called *nonparametric methods*. These methods are usually mentioned in undergraduate statistics courses and a few courses, usually at an advanced undergraduate level, devote 10 or more lectures to them. At the lower end of this 'treatment' scale a textbook is not called for; the topic being covered by one chapter in several general texts. At the upper end, in 1980 there were several available texts, but these tended to be fairly comprehensive and to spill over to the postgraduate level.

At that time there was an increasing interest in nonparametric methods among research workers in fields other than statistics, many of whom were 'do-it-yourself' statisticians. I planned and wrote a book that described the methods in a not-too-mathematical way, with emphasis on practical applications. I wanted to help research workers, and at the same time provide useful background reading, or even a text for, students exposed to a moderately short course on nonparametric methods. Despite good reviews (and a well designed, appropriate, and striking cover), sales were disappointing. With hindsight, I realized the book fell between two stools. I had not thought clearly enough about what my target readership would want. I did not give the 'do-it-yourself' statisticians enough detail, forcing them to look at more formal texts for specifics. At the same time, by covering a broad spectrum of ideas rather superficially, it became too diffuse for most course or examination needs.

I decided there might be a larger readership for a more formal, yet still elementary, book that dealt with key topics more fully but cut back on generalities. I hoped this would fulfil both the role of an elementary student text and that of a guide for research workers in other fields. This time both reviews and sales were encouraging.

This second book fell in that grey area between a formal textbook and one for background reading. It included worked illustrative examples (often features both of a textbook and of one for background reading) as well as exercises for students (primarily a textbook feature). There were tables and figures; appropriateness of these in a book for background reading will depend upon the subject matter. Some basic material of a general statistical nature — which would be familiar to undergraduate statistics students but not to all research workers in other fields — was put in an appendix. I included over 200 references to other books and research papers. I doubt whether any one reader has looked at more than half a dozen of these, but I was confident that most of the references could prove helpful to at least one reader.

A second, completely rewritten, edition of the book was needed within four years of original publication because rapid changes in the attitude towards nonparametric methods took place in the early 1990s — due largely to a breakthrough in computer software. This made tests, once ruled out in practice because they took hours or even days to perform, feasible because they now took only minutes, some only seconds to carry out. This is a good illustration of my point in section 7.3.1 that new developments may create the need for a new textbook.

Readers of background material tend to be critical; they are often the 'better' students, nonspecialists needing information for their work, or the enthusiastic layperson. Never talk down to such readers; there is a danger of doing this unintentionally while trying to give a simplified explanation of a complicated idea. Get facts correct; some of your readers (and reviewers) will be only too happy to point out inaccuracies. Avoid phrases like *Obviously* or *It is easy to see that* . . . unless you are sure your target readers will agree. What might be easy for you as an expert, may not be easy for your readers. The more successful the book, the more you will suffer for such a lapse, being inundated by requests for explanations of why *It is easy to see that* . . . .

Another kind of background reading is what I call a 'link' book. For example, the chemical industry is increasingly under pressure over alleged environmental damage by chemicals. A book highlighting environmental implications of industrial practices with chemicals might find favour as background reading for chemistry students (perhaps also for executives in the chemical industry!) and it may also appeal to environmentalists. If you want to write a book like this, I must warn you that there are books covering the field I chose for this example; but these date quickly as our knowledge of pollution problems grows; at the time of writing acid rain is not so high in the popularity stakes as it was some five years ago; disposal of toxic chemicals and of nuclear waste are current front runners. It may be something else by the time you read this.

7.4 SCHOOL TEXTBOOKS

If you are writing a school textbook the odds are that you are, or have been, a school teacher. This is a specialized field; your book must show the skills needed to hold pupils' attention. In this you may be aided by modern book design, especially attractive layout.

A glance at most recent school textbooks — whatever the subject — will show that nearly all pages have broken text. The breaks may be formed by pictures, diagrams, boxed material, tables, listings and displays. Colour printing is becoming more common, although the economics of publishing mean this is only likely to be used extensively in a textbook if high sales are anticipated.

Breaking up material to hold pupils' interest is a skilled job. Photographs, sketches or diagrams should be relevant and be used in a way that will impress pupils. This may mean they convey information in the most digestible form, or else that they carry a message. There may be an element of almost-slick presentation, calling for the use of the right tool in the right place. If you are writing a geography book, the chapter on Brazil might include a colour picture of Rio de Janeiro harbour and the famous Sugarloaf. However, if you are writing a book on geology, such a photograph would be almost an irrelevance in your chapter on Brazil, for Rio harbour, spectacular as it is, is hardly relevant to that country's rock formation. A picture of some of the rugged mountains behind that city may be more appropriate for a geology book.

At a lower display level, highlighting (or shading) formulae in an arithmetic book should be confined to those that are important; perhaps only to those that students are expected to remember. Over-using a feature like highlighting dulls the impact, but be consistent in the level at which you use it. If you restrict it to formulae that you think pupils should remember use it for all such formulae, but not for others.

'Boxes' are useful for supplementary material to break up the main flow of text. For example, if your textbook is a guide to modern

---

TEXT BOXES

Text boxes are prominent features in many textbooks, where they are used partly to break up pages. In this box you will see that I have used a different font to that in the main text. This paragraph is 'left justified'.

As a contrast in this paragraph the text is 'right justified'. I have done this only for illustrative purposes. The effect is not pleasing.

---

word-processing and your introductory chapter is about the main tasks that can be carried out by modern word-processing software you might include in a box on the first page a list of the five best-selling word-processing packages in the USA (or in the world). Putting this in a box rather than burying it as a list in the main text is useful for the pupil who wants to refer back quickly to the names of the top performers. I have put an almost trivial example of use of a box at the top right of this page.

Unless you are preparing CRC or manuscript for typesetting from disc, ask your publisher how you are to present material to be put in boxes, or matter to be displayed or otherwise integrated with text. Your editor should also indicate or be asked what devices may be used; e.g. whether colour photographs are allowed. Your contract may stipulate that you are to provide exercises and if so how many, and whether they should be integrated with the text, e.g. dispersed throughout at appropriate points, or collected together at the end of each chapter. Decisions on such matters may depend upon whether the book is to be free-standing, or is to be one of a series for which a house-style has already been specified. Some publishers of school texts have even more rigid rules about style than do the publishers of self-teaching books.

In your main text you may find it useful to include summary lists at appropriate points. For example in a geology textbook that has a chapter on field studies in the Northern Highlands of Scotland, you might open that chapter with:

After reading this chapter you should know:

- the names of sites of particular interest to geologists;
- the main rock types encountered;
- the role of glacial activity in shaping the terrain of the Scottish Highlands.

In most countries some science is now taught in primary schools — where there are very special writing problems with textbooks, e.g. matching language to age — but most scientific and technical texts for schools are at the secondary level. At that level strive for clarity without appearing to talk down to your reader. It helps to achieve this, as it does for all writing, if you avoid long words when a short one will do, e.g. *begin* or *start* rather than *commence*; but do not fall into the trap of using an inappropriate word because it seems simpler; this gives immediately an impression of talking down, e.g. when writing for secondary-school pupils, say *your paternal grandfather* rather than *your dad's dad*. I would also avoid *your father's father* in this case. Although this is not talking down, and seems simpler than *paternal grandfather*, the double use of *father* makes it harder to grasp quickly the precise relationship; but this is a point on which opinion may well differ.

In textbooks strive for realism, even if you have to simplify reality, in examples and exercises. As I pointed out in section 6.6, school pupils, like older students, are quickly bored by unrealistic examples.

An important aspect of writing school textbooks is targeting your readership. The school syllabuses in the UK are becoming increasingly constrained by the growth of national curricula as well as by examinations such as the General Certificate of Secondary Education (GCSE) and the 'A'-levels or their Scottish equivalents. This means there is an increasing tendency for school texts to be targeted at a specified syllabus. The same is now true in many other countries.

Where student exercises are included you must decide whether to provide answers; and if you decide to do so, how detailed these should be. In some subjects you may be wise, or even be required by your contract, to write a companion 'key' volume to be supplied only to teachers; this will give detailed hints and solutions to exercises.

7.5 POPULAR SCIENTIFIC AND TECHNICAL BOOKS

Successful writers in this field are often enthusiasts — about their subject and about writing. This dual enthusiasm is a good start, but a further key to success is balancing one's zeal. Too much fervour for your subject may lead you to underestimate the need to explain what, to you, seems obvious. Excessive keenness for writing *per se* may tempt you to give too superficial a treatment in the hope of attracting a wider readership.

If there is a key ingredient of successful popular writing, I am afraid I cannot tell you what it is. A novel approach to your subject helps; timing may also be important. Stephen Hawking's *A Brief History of Time* is no easy read; I suspect that had it been written ten years earlier it would not have made the impact it did. When it appeared there was, I believe, a growing interest in the origins of the universe and how it has evolved. Similarly Roger Penrose's *The Emperor's New Mind*, appeared at a time of increasing interest in artificial intelligence; it attracted readers also, I think, because it presented both a new and a controversial viewpoint. I doubt whether the author of either book made a conscious decision about timing; more likely it was partly a case of the time being ripe for such books, combined with their being well written, that resulted in their success on the popular market. Timing decisions are probably more consciously taken by those writing, for example, political memoirs, or even a formal textbook.

Both the above-mentioned popular books tapped the upper end of that market. Neither is easy reading and both appeal essentially to people seriously interested in their subject matter. Many popular scientific and technical books succeed because they put the public right on topics about which they feel some unease. In my own subject of statistics, a good example is Darrell Huff's *How to Lie with Statistics* (FRIII), for many years a best-seller on both sides of the Atlantic. Its lucid exposure of the tricks of advertisers, politicians and the media in manipulating data to mislead people not only entertained, but played a valuable role in teaching people how to detect this kind of deception. Huff did not invoke high-powered science to do this.

It is possible to write series of scientific best-sellers based on a theme. Martin Gardner's many books on mathematical puzzles are good examples. The mathematical conundrum — like the crossword puzzle — has a wide popular appeal as a mild, often slightly frustrating and irritating, mental exercise. Books on games of skill such as bridge or chess are also, in a sense, popular scientific writing.

For some 50 years, up to the 1990s, many popular books on nearly all aspects of science (and other topics) were published by Penguin books under their Pelican imprint. The demise of that series, with its well-defined format and aims, indicates changing fashions in attitudes towards popular science — and indeed in subjects on a broader front — rather than a loss of interest in these topics. One factor inducing change may be an increase in the space and time devoted to science in serious newspapers and on specialist television channels.

If you aspire to write for the layperson or for scientists in other disciplines, count yourself lucky if you attain best-seller status; but many popular books written by academics, scientists and technical experts in industry, or by journalists with a scientific background, achieve respectable sales of between ten thousand and fifty thousand copies in paperback. Novelty, topicality, lucidity, readability (avoid all but essential jargon and technicalities) all help, providing your writing is informative and accurate.

If you write for the 'popular' market you may have to decide between an impartial approach or taking one side if you choose a controversial issue. Environmental issues are one field where such decisions may be needed. A savage attack upon those who pollute the atmosphere with sulphur compounds, CFCs and oxides of nitrogen will win the approval of the green lobby, though that alone may not be enough to encourage environmentalists to buy your book. Unless it contains new evidence, or makes a new and convincing case for action, environmentalists may take the line that they know all this anyway — so what? Similarly, if you make the case for industry to do little about these problems because the costs of preventive measures do not make economic sense, you will certainly not attract the environmentalists; yet industrialists may also be lukewarm about your book, regarding it only as confirmation of what they already believe, or accept as gospel. If your arguments are weak, readers to whom you are appealing may condemn the book, fearing it could produce a backlash from those holding contrary views. If you give an unbiased account of an environmental issue there is a danger of appearing rather dull and again not winning strong support from either side. If you are keen to write on such a topic do not be put off by my negative comments. I am only pointing out potential pitfalls. You may feel able, and be able, to avoid them all.

On other issues where people tend to have strong views, the question of whether to take sides or try and be impartial again arises. Does one pander to the anti-blood-sports and animal-rights lobbies by highlighting the more questionable aspects of minority sports like fox-hunting or grouse-shooting, or should one try and make an assessment of the value of these activities in maintaining a natural balance and as important facets of a tourist industry in areas that have limited alternative sources of income? Again, what about the use of live animals in laboratory experiments? Your personal views may determine the line you take on unbiased versus biased approaches.

When writing for a popular market be careful to get the right level of rigour in your approach. Avoid inaccurate or ambiguous statements, but this is not the same as omitting information that, although accurate, is only marginally relevant to a point you are making. For example, if you are writing about the environment it may be a fact that per head of population, sulphur dioxide emissions on some remote island are many times the international average, making them the highest in the world. This might be because the island happens to have (a) an iron foundry and (b) a small population. Yet this may be virtually irrelevant to international problems with acid rain (i) because the *total* emission from this island is small and (ii) because the island is so remote that the effect upon rainfall in other countries is too small to be measured.

In popular writing, do not be afraid to use legitimate journalistic devices to surprise your reader. One such device is to explode a popular misconception. In Sprent (1988, p.21) I pointed out:

Nearly 2000 pedestrians die on British roads each year; it is an exceptional year if commercial aviation air fatalities reach that number worldwide.

I used this factual information to introduce a discussion about people's perception of risk. Most people are content to be pedestrians, but many regard flying as hazardous. Be careful though; do not over-use any one trick such as surprise or highlighting unexpected truths; the novelty wears off.

When seeking a publisher for a popular book remember that these may appeal more to mainstream than to academic publishers, and your chances of getting an advance on royalties should be better than they are for specialist academic or technical writing. If you are writing for a popular market you may find an unexpected interest in translation rights, particularly in topics such as the environment, nuclear power, etc., that may be of great interest to readers in some non-English speaking countries. Look carefully at any clause in your publisher's contract about your share of sums received for translation rights. While 60% may be acceptable for academic books, look for a higher rate for popular writing.

One point that worries some writers, but upon which I have no clear view, is the use of humour in basically serious writing. Relevant humour was an essential ingredient in Huff's *How to Lie with Statistics* and most readers will enjoy a good laugh. There are two difficulties with humour.

- What may be funny to some readers may seem downright boring, or in bad taste, to others.
- Some humour — especially any depending upon local idiom or a national characteristic or event — may be completely meaningless to readers in another country.

My advice is to bear these difficulties in mind if you want, or are tempted, to be funny. If you are happy that humour is relevant, or necessary (as it was in Huff's case), or that your readers will enjoy a moment of light relief go ahead and use it; but choose examples that are likely to be understood and appreciated internationally if you are aiming for export sales. Avoid forced humour or any 'formula' for introducing it (e.g. do not make a rule that you will have at least one funny remark in every ten pages). Even a reader who does not twig your formula will realize something peculiar is going on. Readers find nothing funny about an author trying to be funny and not succeeding; that is the usual fate of forced humour.

An example of the sort of humour I have in mind to provide light relief would be to insert in an article on telephone accessories a line like:

A telephone answering machine is a useful device for letting potential burglars know how long you are going to be out.

Maybe you do not think that is funny. I tried it on a dozen or so people; some thought it funny, some thought it 'corny'. One didn't know what a telephone answering machine was.

8

Other work-related writing

8.1 A WIDE RANGE

Scientists and technologists who do not have, or want, to write a book, still have to communicate. You may have to produce an instruction manual, prepare an expository paper or a research or planning report for general release or for internal company use; these are usually *have-to-do* tasks. Writing popular and informative papers, or describing new developments in trade or professional journals aimed at broad groups such as engineers, computer scientists, the pharmaceutical industry or management are other common jobs. You may have to write reports for in-house circulation, for publication in company bulletins, or for trade or industrial group magazines (this is sometimes *have-to-do*, sometimes *want-to-do* writing). Experts may be asked to write advertising copy for those who need, or want, technical details. Reviewing books in one's field of expertise is another task — one not always as well done as it deserves to be. This chapter covers such activities.

The writing I look at in this and the next chapter differs from book writing in matters of style and also in commercial aspects. Payment is often at a flat rate ranging from nothing (if the writing is part of your job) to £100 or more per thousand words. Contractual arrangements range from formal written agreements to loose verbal ones. If there is a verbal agreement involving payment, insist on written confirmation of the main terms.

I cover instruction and other manuals in section 8.2. Writing one of these is sometimes as big a task as writing a book. There is a growing demand for them stimulated partly by new technical developments and especially by advances in computer software. Writing a manual is an in-house job in many companies, but there is an expanding market for freelances with the right technical know-how and the all-too-rare ability to frame instructions clearly, for clarity is an essential ingredient of all good instruction guides.

8.2 INSTRUCTION MANUALS

8.2.1 Users' background knowledge

Instruction manuals are needed for lawn-mowers, cookers, refrigerators, vacuum cleaners, washing machines, tumble dryers, cameras, cars, computer hardware and software, hi-fi equipment and dozens of other technical products. If you have never been irritated or confused by one of these manuals you are either lucky, well versed in technical matters, or easy to please. I say this despite a noticeable and welcome improvement recently in manuals from leading manufacturers of consumer goods. Alas, this is faint praise, meaning only that many manuals are now reasonable, whereas once the majority were bad. To what extent a good manual helps sales is unclear, but a bad one certainly discourages customer recommendations — a well-known sales generator. There is evidence, confirmed in 1993 by a survey carried out by Dell Computer Corporation, that a prime cause of technophobia — fear of modern technology — is an inability to understand instruction manuals.

A writer of a manual for any but simple equipment must always remember that users are likely to have anything from zero understanding of how the item operates, to knowing most of what is relevant to its efficient use. Dissatisfaction with manuals often stems from writers failing to come to grips with this problem. A common mistake is not distinguishing between know-how needed to make best use of a product and technical information about its construction. Much of the latter is irrelevant for a user, and should usually be left out of instruction manuals. Writers sometimes include it because — perhaps subconsciously — they want to show how much they know.

For example, using certain insulating materials, air-circulation systems and positioning of heating elements may ensure that an oven maintains a uniform temperature throughout. To know that it does may be important to some users, but they do not need to be given technical details about how this is achieved. The manual need state only that the temperature will be uniform throughout the oven (if this is true!) and perhaps even more importantly, if the temperature is not uniform, to indicate how it varies. If the temperature on the lower shelf will consistently be 10° C lower than that on the upper shelf, and the temperature setting specifies the temperature for the upper shelf, this may be important for some users, especially if the oven is for laboratory, rather than domestic, use.

If new technologies mean that users must change or adapt conventional approaches this should be highlighted in a manual. For example, microwave ovens use a different mode of heating to conventional ovens and many housewives — and househusbands no doubt — have problems with food boiling over, or coming out scalding hot on the inside and cold on the outside, even when they have followed cooking-time instructions to the letter. These problems are often due to using an unsuitable container or inappropriate wrapping. Good microwave instruction manuals warn of such problems and list the main causes; if your manual does not, it is a poor manual. A trained physicist who understands the principles of microwave cooking would foresee most of these problems, but most domestic users are not trained physicists.

The best known and best selling genre of instruction manuals are household cookery books. These deal with particular skills rather than specific equipment and because the market is so large, their writers can, and usually do, target subsets of potential readers. These range from those concerned only with the basic needs of eating to live, to the diverse demands of those who live to eat; the latter is a market that includes professional chefs, skilled amateur cooks and struggling enthusiasts. If you have aspirations to, or are required to, write any sort of instruction manual — even if the subject be far removed from the culinary arts — it is worth looking at a range of popular cookery books to see the different ways their authors approach their task. Most of them target a specific sub-group of the total market, something you are usually not able to do, except in a limited way, when you are writing a manual for users of a sophisticated lawn-mower or the latest computer software.

Depending on your own culinary skills, you may find some cookery books boring; this may be because the recipes seem dull and so banal that they are hardly worth describing, yet are given in great detail. At the other extreme, you may be baffled by descriptions of exotic foods and obscure instructions for preparing and mixing ingredients.

A competent amateur cook will probably have no trouble understanding the following instructions which I found (among others) in just one recipe in a widely-used general cookery book.

- Place the lemon peel and cinnamon in a saucepan.
- . . . then put through a fine strainer.
- Mix the liquid, when slightly cooled, with the flour.
- When smooth place in a saucepan.
- Cook over moderate heat.
- Add a walnut of butter.
- Drop a dessertspoonful of this batter into hot lard.

- Add a few at a time.
- Fry till golden.

Since my culinary skills are limited to boiling water without burning it, or obeying explicit instructions like *heat can in boiling water for 20 minutes*, I need guidance on each item in the above list. My queries, in order, on each of the points above would be as follows.

- How large a saucepan?
- How fine a strainer?
- How cool is *slightly cooled*?
- How smooth?
- What is meant by moderate?
- What is a *walnut* of butter?
- How hot?
- How many is *a few*?
- Is *golden* a light yellow or a deep rich yellow-brown?

Clearly, if I want to widen my culinary horizons, I would not get on well with the book where I found this example. This is not the author's fault, for she (it was a lady) was clearly aiming at a target readership with aspirations and experience greater than mine.

You may think this a trivial example of the obvious, but problems in imparting information about how to do things, or use equipment, increase with sophistication. While most computer users need know little about how a silicon chip functions, they must know how their machine will respond to each instruction. Computer buffs know most of the answers, but computer buffs are only a small percentage of those who use computer software for business or scientific and technical applications. The relevant points must be explained for a beginner who does not understand computer basics, and for whom most computer jargon is meaningless. Many computer manuals fail to appreciate the latter point.

8.2.2 A range of uses

Equipment or utilities often have a range of features, not all of which interest, or are needed by, every user. This problem faces writers of instruction manuals for computer software packages for, say, word-processing. For example, a facility for merging documents will be widely used by a company that sends similar, yet personalized, letters to many people on a large mailing list, for it enables them to match names and addresses from a list (one document) to the 'common' letter text (another document). But that facility may be of little interest to a scientist who uses the software mainly to write papers and reports, or to prepare lecture notes and student tutorial exercises, although a few teachers or administrators may find it useful, for example, in keeping records of students' performances. Scientists and technologists are more likely to be interested, depending on their subject, in how to incorporate graphs and tables in text. A solicitor's main concern might be with facilities to lay out documents in approved legalistic styles, while the compiler of a company newsletter will want to know how to incorporate the company's logo and the way to present text in columns under a banner headline and with text wrapped around illustrations. Here the word *wrapped* is a technical term that should be explained in the relevant section of a manual. Its meaning is almost — but not quite — what one might expect from the everyday use of the word.

Authors may find automatic indexing facilities useful and also the ability to move large sections of text within files or from one file to another (I often do!). *File* is another piece of computer jargon that will need explaining for the benefit of users new to computing, though as an author or intending author, you should know the meaning in this context already if you use a word-processor.

Although instruction manuals for computer software are improving, some leading software houses might well wonder why, when they produce updated versions of major packages (which may sell hundreds of thousands, even millions, of copies worldwide), there is often a near-flood of books from independent authors purporting to be easy-to-understand guides to that new package.

8.2.3 Computer software manuals

Because they are common, and also because they highlight many problems in manual preparation, I have chosen computer software manuals to illustrate points about manual writing. Different needs, and varying expertise among users of software packages, lead to layout problems. One approach, used by some leading software houses, is to produce several manuals. A major word-processing or spreadsheet package might include the following.

- A beginners' guide (often including a tutorial to give the user 'hands-on' experience with the program).
- A workbook, or applications guide, that devotes sections or chapters to each major use, illustrating these with examples to give further 'hands-on' experience.

- A reference manual which explains available facilities; is a guide to trouble-shooting; deals with importing from, and exporting files to, other programs; linkages to printers and other peripheral devices. Sensible ordering of material and good indexing and cross-referencing is essential for this type of manual; it is going to be the main source of information, apart perhaps from *on-line 'help'* (yes, more jargon!), for both experienced and new users when they run into a problem, or want to do something outwith their usual routine.
- Additional supplementary information such as a pamphlet detailing how to set up the system on various hardware configurations. If the package is an updating of an established program, there will probably also be a list of new features and any changes in operational procedures, e.g. if the old program required you to press the function key **F3** on an IBM-PC keyboard to obtain the 'help' menu, and the new program requires you to press **F1**, most users will appreciate prominent notification of this change without having to search for it in some obscure hiding place.

If you have to write one of these manuals how might you set about the job? On an in-house project you may be constrained by company rules about how its manuals are laid out; even so, if you are sufficiently senior, and you think the management have got it wrong, speak out. If you are a freelance, you may still have to comply with some company rules, but the odds are that if a firm contracts out this work it is because it has not got the internal expertise to do it, and you are likely to have a reasonably free hand.

To make specific points, I consider preparation of manuals for a typical word-processing software package, assuming there are to be three basic documents:

- a beginners' guide;
- an applications workbook;
- a technical reference manual.

Although common, this split is not appropriate for all packages. Specialist software for, say, graphics or statistical analyses, often has only one manual.

First, the beginners' guide. What is a beginner? Somebody new to computing, or new simply to word-processing? The safe answer for a manual writer is 'probably both', because word-processing is often a user's introduction to computing. Your guide should include an account of basic computer concepts and explain computer jargon relevant to word-processing. In catering for beginners beware of a lurking danger of seeming to talk down to more experienced users.

Avoid gratuitous and irritating soft-soap like

Congratulations on your purchase of the world's premier word-processing package.

This helps nobody to use it. It may grate with a beginner who finds mastering instructions hard work and who will probably take the claim with a pinch of salt anyway. It will sound patronizing to knowledgeable users who have purchased the package after careful assessment of the market. I have a gut feeling that this type of soft-soap is less unacceptable to American users than it is to many in the UK; it certainly seems to be more frequently used in American manuals. Avoid, too, pompous instructions or warnings like:

Depression of the escape key will result in a premature cessation of the printout routine prior to completion,

when you mean

Pressing 'escape' will stop printing before it is finished.

Some instruction manuals are noted for long-windedness.

The common division into three, or sometimes more, manuals has evolved from experience in tackling the joint problems of varied backgrounds and differing user-requirements.

The beginners' guide should feature basic applications, slanted towards business use, because this is the main application field for any general word-processing package. That emphasis may not appeal to you, but accept it because it will suit most beginners. Explaining how to write a letter is an obvious early example to use, but first say something **relevant** about how computers work. With almost universal computer training in schools, you may think this is unnecessary in a manual for a sophisticated package, but remember, there are many who left school some years ago and who are now transferring from electronic, even mechanical, typewriters to computers.

A good starting point might be to explain differences between a typewriter keyboard and that for a computer; explain the **Ctrl**, **Alt** and function keys (**F1, F2, F3, . . .**). Mention scroll keys like **→**, **←**, **Home**, **End**, **PgUp**, **PgDn**. Explain the use of 'double keystrokes' such as [**Ctrl, W**] which is to be interpreted as *press first the control key* (**Ctrl**) *and while it is held down, press also the* **W** *key.*

You will need also to describe operational differences; e.g. with a mechanical typewriter, the typist activates carriage return at the end of a line and 'feeds', or turns, the paper to a new line. This is done automatically in word-processing and it is only if you want to 'force' a new line, e.g. at the end of a paragraph, that you need to press the key marked **Enter** (or sometimes just designated with the icon ↵). You will have also at an early stage to explain computer jargon (e.g. *hard disc, disc drive, RAM* and *ROM*) essential for using the package. If there is no separate instruction manual or pamphlet about installing the program on a computer, this should be covered in the beginners' guide.

Explain clearly how you are going to distinguish between an instruction to press the key marked **Enter** and the use of 'enter' as an instruction to do what follows, e.g. 'enter the date'. I once met the horror *to exit enter enter*! There are various ways around this and similar problems. One is to write keying instructions in bold, e.g. **Enter** means press the 'enter' key. A clear convention is also needed to distinguish between double keystrokes such as pressing **Alt** followed immediately by **F1** while **Alt** is still depressed, and consecutive keystrokes such as **F4** followed by **Esc** after **F4** is released; also to distinguish between the function key **F4** and the key sequence **F** followed by **4** (the usual convention is to indicate the former by **F4** and the latter by **F, 4**). I omit details of several widely-used alternative conventions in computer manuals to deal with these problems; it is essentially a matter of defining conventions and sticking to them; e.g. if you decide that double keystrokes are always to be placed in square brackets, with a comma between the designated keys, do this consistently, e.g. always use [**Ctrl,F4**] when you mean the function key **F4** is to be pressed while the **Ctrl** key is held down. If you are using bold to indicate keystrokes, do this consistently and avoid, or at least be careful to avoid ambiguity with, any other uses of bold.

The first chapter of a beginners' guide might appropriately be devoted to considerations like those just outlined. Make the presentation orderly but concise, because readers will be keen to start using the program. It helps, even at this early stage, if simple exercises — involving hands-on experience — can be included to demonstrate, say, key usage.

The next chapter should be targeted at practical letter writing. If you are familiar with word-processing you will know some sophisticated short cuts, e.g. using a 'macro' to produce, with a couple of keystrokes or 'mouse' clicks, your firm's name and address together with the date as a letter header. A beginners' manual is no place to introduce such sophistication. A novice reader may not known what a 'macro' is, and will certainly find it hard to set one up. It is better to ask the reader to use the program to prepare a letter in a way that follows typewriter procedures as closely as possible. Here you will face problems with the order of presentation. What about setting the margins; specifying the paper size? Most word-processors have 'default' settings for paper size — usually European A4 or American letter size — with reasonable (perhaps 1 in. or 2.5 cm) margins at top, bottom and sides. My inclination is to accept these 'default' settings in this first exercise. Explaining how to change margins, paper size, etc., can come after the letter is written. Your manual might give a specimen letter to be typed. You might first explain how to place the address at the top right using the **tab** key to position this correctly. This is conceptually simpler, and more akin to the way it is done on a typewriter, than temporarily resetting the left margin. You must guide the reader through the rest of the typing process to reproduce the draft letter (best shown in the manual as a facsimile of a typed letter) like that in Figure 8.1.

You must explain at an early stage facilities for 'insertion' and 'deletion' of material, because mistakes are going to be made. One approach is to make the letter short (like that in Figure 8.1) and to advise the user to ignore typing mistakes until the letter is completed. You then explain simple *deletions* and *insertions* in sufficient detail to correct typing errors.

The next important step is to explain how to save the letter in a named file. You will have to give the conventions for naming files if this has not been done in the introduction. Saving this letter to disc is important if your manual proceeds the way I have in mind, for it will be used in later chapters to explain features like changing the margins, altering fonts and point size, making major additions or deletions (e.g. adding or removing a paragraph).

You will face dilemmas. When do you introduce printing? Typing onto the screen is fine, especially if the package features 'wysiwyg', the acronym for *what you see is what you get*, but not all packages have this feature and even among those who claim it, few have 100% *wysiwyg*. Even if your package has good *wysiwig*, most people still like to see printed or 'hard' copies of their work. If you are sending a real letter that is your goal!

There are hundreds of different printers in use and 'interfacing' a particular printer to a word-processing package is technically complex. The installation instructions for the package will not describe the technicalities, but should explain what a user has to do to 'install' a printer. If no printer is installed, and you want the user to print the letter, you will have to indicate in your manual at this stage the procedure for assigning a printer, or else state where the information to do so may be found (likely to be either in the installation instructions or in the general reference manual). Give such a reference even if it is reasonable to assume printer installation has already been done at the setting-up stage, because, for all sorts of reasons, various steps in the installation may not have been carried out in the conventional way. In most cases, once a printer is designated and the necessary hardware connections are in place, it only requires a few keystrokes to print a letter.

```
                                    Megaton House,
                                    12 Market Street,
                                    KEYTOWN,
                                    Wessex,
                                    KE8 2BG.

                                    27 February 1993.


The Manager,
Willow Catering Ltd.,
17 Back Rd.,
KEYTOWN, KE1 5DG.

Dear Sir,

We acknowledge receipt of your letter dated 24 February
asking for a quotation for the supply of various items
of cutlery.

It is our pleasure to submit the enclosed quotation and
we hope this will meet with your approval.

Yours faithfully,



J. Bloggs,
Manager,
Megaton Catering Equipment Co. PLC.
```

Figure 8.1 A specimen letter suitable for a beginner's guide to a word-processing package.

You may decide, however, to encourage the user to learn about features to change the format of the letter before giving printing instructions.

Once you have used the letter as a testing ground to illustrate the array of basic formatting changes — margins, line spacing, font types and font sizes, indenting, various screen displays, probably the use of underlining, italic or bold, deletions, insertions and moving text, you will have to consider what other features to introduce in a beginners' handbook. Preparation of tables is a possible choice. At this level a 'hands-on' tutorial would be useful. This could be of the 'interactive' kind; a file with a table or tables might be included with the package and the manual would then indicate procedures for retrieving this file so that the user could alter the basic table — perhaps introducing horizontal and vertical rules, changing the layout of the heading, realigning columns, changing row and column headings from roman to *italic*, adding a footnote. A complete description of all keystrokes for these changes will have to be given, and the manual should include printouts (usually best in 'screen' format) to show the effect of any such changes, so that the user can see if he or she has produced the intended effect by following your instructions.

Other topics to be dealt with in this beginners' guide may be dictated by the markets at which the package is aimed. It may include the creation of graphs and diagrams or the preparation of publicity material such as a news-sheet, which might involve arranging text in columns. Another candidate for inclusion would be information on merging documents, e.g. the afore-mentioned example of adding names and addresses from a mailing list (held in one file) to a standard letter (held in another file). There might also be an introduction to facilities like automatic indexing or rearranging lists in alphabetic order. There should certainly be accounts of how to use tools such as the spelling checker and thesaurus, and sometimes a grammar checker, that are featured in most word-processing programs. My inclination would be to omit from a beginners' manual more sophisticated features like presenting mathematical formulae or preparation of 'macros'; the latter are in essence sub-programs that a user writes to do specific tasks likely to be repeated very often; each may involve hundreds of keystrokes, but by storing these in a macro they can be activated with only one or two strokes.

I have assumed so far, apart from a passing reference to clicking a 'mouse', that all operations are being performed using a keyboard. A 'mouse' (another piece of jargon) is now often used to perform many operations that once could be done only by keystroke combinations. If you are writing an introduction for mouse users, alterations will be needed in your manual. Manuals for major packages often indicate both keystroke combinations and the 'mouse clicking' procedures to achieve the same results; one set of procedures usually being given in a different typeface to the other, or with icons (representing a mouse or a keyboard) to highlight each, beside the alternative instructions. Such dual presentations call for increased concentration from the reader, but I see no better way round this problem. Both sets are needed by mouse users, for often, in practice, they use a mixture of keystrokes and mouse clicks, and so need to know the corresponding requirements. If you are a mouse user and operate in *Windows*, you will know what I mean when I say it is sometimes easier to press **Enter** than to position the mouse and left-click the **OK** button in a dialogue box. That last sentence has a nice collection of jargon that would need explaining in your manual — *position the mouse, left-click, OK button, dialogue box* — just a small taste of the jargon associated with 'mouse' technology and the *Windows* (more jargon!) operating system.

The workbook, or applications guide, will differ from the beginners' guide in that it covers more sophisticated aspects of each item discussed. However, material should still be arranged by type of application. It is reasonable to assume the reader has studied the beginners' guide, and to give references to topics already described there in detail, rather than to repeat such material.

You may decide to split the workbook into two parts; the first dealing with basic applications (letter writing, preparing newsletters, reports and legal documents, creating tables and diagrams, making an index, etc.) The second part would then include instruction on topics like constructing and using macros, importing files from, and exporting files to, other programs, restructuring files (e.g. rearranging items in lists in alphabetic order, etc.) and any special features such as graphic facilities.

It is an open question whether a widely used but fairly technical facility like that for merging documents (used not only for sending a standard letter to a number of different recipients on a mailing list but for combining information or records from a number of different sources) would be described in the first or the second part. My inclination would be to put it in the second part, but that may be because it is a facility I seldom use. I would certainly put in the second, rather than the first part, instructions on how to alter detailed settings for tabs, sophisticated ways of aligning figures in tables, use of 'kerning' in printing, and details of how to alter the spacing between letters and words or to introduce additional fonts or incorporate special characters such as Greek characters or mathematical symbols, all things which can be done with most word-processing packages.

Some discussion of sophisticated file management would also be appropriate in this second part. File management covers naming of files, their allocation to directories so that they may be rapidly accessed, the making of back-up copies of files (and any manual that does not stress the importance of making these is falling down badly on its job), re-naming of files, deleting them, combining them and moving material within a file and from one file to another. For an author, particularly one preparing CRC or discs for typesetting, the facilities for transferring material from one part of a file to another (e.g. moving tables and diagrams) are especially important. So is the ability to transfer material between files. For example, typically when working on a book each chapter will form a separate file, but authors often want to transfer a paragraph or even a whole section from one chapter to another. Part of Chapter 6 of this book was originally in my draft Chapter 2. I did not want to, and I did not have to, retype it using the keyboard.

Clearly the precise content and layout of a workbook will depend on the package and the market it is targeting. While general usage word-processing packages are aimed primarily at the business user, they also provide facilities of special relevance to authors, scientists, etc. There are however one or two more specialized packages designed specifically, for example, for authors and journalists or for technical users. The general warning about keeping in mind the different background knowledge of users applies particularly while writing, usually detailed, applications workbooks. Although these manuals should give reasonably self-contained accounts of each topic covered, it is quite in order, indeed virtually essential, to give references to a beginners' guide or to a general reference manual for further details, especially if these details are likely to be relevant to only a few users.

The third, or reference, manual will normally have a different layout, dealing in full with one function of the package at a time. It will not explain how to write a letter or prepare a legal document. Instead, it will describe the available features or tools in the package, perhaps, but not necessarily, arranging topics in alphabetical order under broad headings like File organization, Default settings, Importing and exporting files, Merging, Assigning printers, Trouble-shooting, Printing functions. Many of these sections will have special aspects best put under prominent sub-headings; for example in the section dealing with printing functions·there would probably be subsections dealing with fonts, with kerning, with producing subscripts and superscripts as in x^2 or x_n . There would also be information about special typefaces such as bold or italic, and about facilities for producing special characters like mathematical symbols or Greek letters.

The reference manual must cater also for features that may only interest a few users. For example, in setting up complex mathematical expressions, normal or 'default' spacing often leads to an unpleasing appearance and one may want to alter the spacing between some words and characters. If the package allows this (and most sophisticated packages do), the reference manual must include instructions on how this is done.

8.2.4 Policy and guidance manuals

In addition to instruction manuals for equipment and specific tasks such as cooking, there are many manuals or handbooks that explain policies or strategies, or indicate how to do a certain job. These are commonly produced by government departments or agencies, by business organizations or by professional bodies. Like instruction manuals for using equipment these should be clearly written and relevant to the target readership. Some of these handbooks are hybrids between a textbook and an instruction manual. The most common criticisms of such documents are that the style is often pretentious and that too much jargon is used. If these documents are intended for wide circulation they are sometimes very elaborately produced, but attractive layout and typography sometimes draw attention to weaknesses in content and style rather than hiding such defects.

In an entertaining, yet serious, article in *The Independent* on 6 May 1994 entitled *In plain English, your jargon makes me sick*, Stephen Pimenoff, for many years a teacher, is highly critical of handbooks issued by some educational institutions and authorities. He rightly condemns the following passage from an Initial Teacher Training programme both for its monotony of style and excessive jargon.

> *Intending teachers need to have a critical awareness of the potential of teaching to promote the processes and realise the factors influencing the acquisition of skills, and an understanding of the effects of previous experience and cognitive limitations in relation to factors of meaningful grasp and deep learning.*

I have not done a very good job if any reader of this book could not improve on this!

I recently received a copy of a lavishly-produced report entitled *Forward Look of Government-funded Science Engineering and Technology, 1994*, a report presented by no less a person than the Chancellor of the Duchy of Lancaster. It has a manual-style layout. It may be quibbling to say I would prefer to take a forward look *at*, rather than *of*, something, so let that pass. The document was glossy and lavishly produced, but I was disappointed by the microstyle and some of the content. It was heartening to find that *Nature* agreed with me in an editorial comment shortly after the document was published in May 1944. That journal sub-headed its comments

> *An over-glossy statement of what the future holds leaves many important questions without answers.*

The *Nature* article asserted that typographically the British Government has done the scientific community a great honour and gave a description of the attractive layout. The writer then turned to more serious matters, saying:

Sadly, the text does not live up to its appearance, either in literacy or in content. The British Civil Service . . . has taken this opportunity to launch the novel noun 'build' which appears to signify the amount of building (physical or metaphorical) accomplished at some defined instant.

There was more in similar vein about style before the writer turned to content, a matter not directly relevant to this book, apart from the suggestion that the document was unlikely to satisfy the declared target readership.

8.3   EXPOSITORY PAPERS AND RESEARCH REPORTS

Publishing research results is a form of *have-to-do* writing that academic scientists and many technologists cannot — indeed do not want to — avoid. Despite readily available advice on how to present material, editors of journals often complain, and I confirm this from my own experience, that much of what is submitted is poorly written. An experienced journal editor, Klaus Hinkelmann (Hinkelmann, 1993) recently wrote:

> The manuscript should be clearly written. . . . authors should not use referees as their first sounding board with the attitude, 'Oh, well, the editor will ask for changes anyway, why should I spend so much time and produce a polished product.' Referees will notice this quickly and put the manuscript at the bottom of a huge file from which it will never emerge, or only after a long time.

How to write expository papers and research reports is a main theme in Maeve O'Connor's *Writing Successfully in Science* (FRI) and in John Kirkman's *Good Style — Writing for Science and Technology* (FRI); also a feature of Kate Turabian's *Manual for Writers of Term Papers, Theses and Dissertations* (FRI) and one covered in Blake and Bly's *The Elements of Technical Writing* (FRI). O'Connor deals with specific points about how and where to submit papers as well as with style, while Kirkman deals in a more detailed way than I do with style, especially microstyle, for this type of writing.

Unlike book writing, or preparing manuals, where controlled redundancy should be used to aid clarity and make for ease of reading, papers for learned or professional journals that report research, or that are expositions of a clearly defined topic, require economy of words. O'Connor's advice is

Aim for clarity, directness and precision.

Style should follow any guidance given by a chosen journal and the content must be appropriate for that publication. The name of a journal often gives a hint of the type of paper it publishes, e.g. *The Journal of Experimental Botany* and *The Journal of Theoretical Biology* clearly cater for papers dealing with different aspects of the biological sciences. Titles may also be misleading. If you are not familiar with it, you might be surprised by the contents of a typical issue of the journal *Biometrika*. Many librarians seem to be, cataloguing it under *biology*. Unless you are already familiar with a journal and its coverage, study recent issues and convince yourself that your paper falls within the scope of that publication before submitting anything; if you do submit, get your paper in the correct style and format. Many journals indicate not only the types of paper they are willing to publish, but specify style and layout requirements which may include *macrostyle* guidance like

Mathematical detail should be kept to a minimum or clearly signposted, with necessary details placed in an appendix.

Instructions to contributors are often given on the inside front-cover or towards the back of each issue of a journal. Read and follow these instructions. If none are given, but you think your paper may be appropriate for that journal, it may be worth writing to the editor or publisher (enclosing a stamped and addressed envelope) asking if specific instructions are available. Most reputable journals will respond, even if the answer is only 'no', or 'follow the style of papers in recent issues'. In many subject areas there are widely used formats for the structure of research reports, used almost irrespective of where the paper is published. A common structure for a paper reporting results of laboratory or field experiments is to have sections headed

- Introduction
- Methods (*or* Materials and methods)
- Results
- Discussion
- Conclusions

but sometimes the Discussion section is omitted or combined with Conclusions. The Introduction is often preceded by a Summary and most journals also require an Abstract, which should be kept free of formulae and equations (mathematical, physical or chemical). Abstracts are often used verbatim by abstracting journals. For an expository paper that summarizes the current state of knowledge on a particular topic, a rather different layout will clearly be needed, although a *Discussion* and a *Conclusions* section are often included. Many journals also require, either after the abstract or as a footnote, a list of keywords. Keep these few in number and relevant to the main theme of your paper.

Follow carefully any instructions about how diagrams and tables are to be presented and numbered. If, as is often the case, you are asked to provide multiple (usually three) copies of your paper, do so. This request is designed to speed up the refereeing process. Photocopies usually suffice for the second and third copies. Some journals accept, or even prefer, submissions on disc with specified formats; a few accept, or ask for, CRC.

When writing for learned or professional journals that publish research papers it is usual to assume that readers will be experts, and unless you are introducing new terminology (and do this only if it is really necessary or you can see positive benefit in doing so, as I did when I introduced *macrostyle* and *microstyle* in this book) there is generally no need to explain jargon. Remember not only O'Connor's advice about *clarity, directness and precision* but check carefully the logic of your arguments and look for gaps in the evidence you use to support any hypotheses. Cite all relevant work. These points are important not only because they help to enhance your standing, but because of the so-called peer review system whereby all submissions to reputable learned journals are sent to referees for assessment. Only if you are an expert in your field (and if you are, you should hardly need the guidance in this section) does peer review mean what it says. If you are new to the field, or indeed writing your first paper, the referees will not be your peers, but people far more experienced than you are.

Some papers, but only a small proportion of those submitted to most journals, are accepted without further ado after refereeing. If there are criticisms of your paper from referees, copies of the referees' reports to the editor, or relevant extracts from these, will usually be sent to you (without revealing any referee's identity) and the editor will indicate either:

(a) that he or she must reject your paper in the light of these comments, and perhaps also for other stated reasons, e.g. that it is 'too theoretical' for that particular journal or
(b) that the editor is prepared to consider a revised version that meets the referees' criticisms.

In the latter case, if the criticisms are important and valid, there is little likelihood of your paper being published by that journal without major revision. Your best option then is to do further work, and if this enables you to overcome the objections, submit a revised version to the same, or another, journal. Usually, take the latter course only if the nature of the revisions is such that the updated paper is clearly no longer appropriate for the first journal.

If a paper has been fairly criticized, resist the temptation to submit it to a second journal without revision in the hope that that journal's referees will not be so critical. If the original criticisms are of some weight, then even if the paper survives the second journal's refereeing process and is published, warts and all, other experts in the field may spot the defects, and this will not help your reputation. It is also possible that the second journal will send your paper to the same referees as did the first one, and if there has been no revision they are unlikely to drop their objections; indeed, they will almost certainly express them more forcefully.

If you can meet valid criticism simply by amending your presentation without having to do more laboratory or field work or develop further theory, then make the necessary changes and resubmit.

Sometimes you may believe that a referee's criticisms of your work are unjustified. You may be right; referees are not God, although a few endow themselves with an unwarranted god-like status. If you can make a reasoned case for refuting a referee's criticism, do so in a politely worded letter to the editor.

Occasionally two referees (most journals use at least two) disagree on the merits of a paper. If the editor feels that a referee who likes your paper has a good case, but that the referee who dislikes it might have some reasonable, yet not compelling grounds for doing so, the editor may invite you to consider the criticisms and to make changes if you so wish. Sometimes you will have to read between the lines of an editor's letter to decide whether the matter is being left entirely to you, or if you are being strongly urged to make what will often be relatively minor changes to overcome the dissenting referee's criticisms.

8.4 IN-HOUSE REPORTS OR ARTICLES FOR TECHNICAL OR TRADE JOURNALS

I discuss these topics under one main heading but in two subsections because both types of writing are aimed at specialists or technically competent people who are not necessarily, indeed often are not, expert in the author's own field. In-house reports are usually written for a clearly defined group such as managers, accountants, production engineers, marketing staff, etc., whereas articles for technical or trade journals will be written for a readership that is largely defined by the type of journal; these might be engineers in general, motor industry engineers, computer systems analysts, retail distributors of machine tools, etc. Journals of this type usually emphasize practical aspects more strongly than do those learned and professional journals I had in mind in the last section that, in the main, report or summarize research results.

8.4.1 In-house reports

In-house reports are usually called for by management. As an employee you may have to prepare a regular monthly or quarterly report on your work or that of your department; or you may have to prepare a one-off report. A report for the accounts department should have a different style and content to one written for the company's engineering advisers. If it is appropriate, you may use financial jargon freely, and without explanation, in a report for accountants, but keep off chemical or engineering jargon unless it is essential. If you must use it, explain it.

Order of material is often important for in-house reports. The readers are likely to be busy people. Your report will be competing for attention with inter-office memos, agendas for meetings, perhaps the latest issue of *Playboy Magazine* and all the other paperwork that constantly harasses managers. If this last sentence did not make you smile, see my comments about humour near the end of section 7.5. For many reports, some readers will be concerned with detail, while others will only be interested in broad conclusions or recommendations. The rules of macrostyle and microstyle given in chapters 2 and 3 apply; use tables, graphs, diagrams or photographs if appropriate. Layout is likely to be largely in your hands; the report will probably be a desktop publishing (DTP) job, using either a DTP page-maker program or a sophisticated word-processing package. A limited circulation report may be produced on an inkjet or laser printer and photocopied. If your report is aimed at selling an idea, or a product, or if it may have important repercusssions on your company's policy, then a well laid out, well presented report may swing a borderline case the way you or your company want it to go.

For some projects you may have to report to two, or occasionally more, disparate readerships. For example, if you are preparing a report on a new machine tool that your company hopes to sell to the construction industry you may have to:

- convince engineers that it will do the job;
- prove to accountants that the costing is right.

The engineers will want technical information, probably including detailed specifications; the accountants, financial data. Jumbling the two in a well-stirred potage will help neither decision-makers nor sales prospects.

You might write two reports; one for accountants and one for engineers. There will be some, perhaps a lot, of overlap. Company policy, or time factors, might rule out this approach. If there is to be only one document signpost it to help readers pinpoint the portions likely to interest them. You might divide the report into sections, starting with an introduction describing the equipment in broad terms, what tasks it will perform, and its advantages over existing tools. Keep this introduction free of technical terms likely to be known only to accountants or only to engineers. The second section might be for engineers and include specifications and operating characteristics; technical terms familiar to engineers may be used without explanation, but if a key feature of your machine is a new component that your company calls the *bifocal neutron sweeper* then explain this in as much detail as is needed for potential users to assess whether it will help them. The final section might be aimed at accountants; here standard financial jargon is acceptable. Don't forget that golden rule — avoid verbosity. Neither engineers nor accountants care whether you know a lot of obscure words; these only distract from clarity. I emphasize this point because verbosity is often the Achilles' heel of technical writing at the report level. Do not write

> The initial working hypothesis which is assumed to hold until further evidence becomes available (when it may be subject to adjustment) is that the output per man hour with this new equipment will be for all practical purposes identical with that attained from the already existing equipment.

when you mean

> Until contrary evidence is found, we assume output per man hour to be almost the same for the new and for the existing equipment.

The latter saves words and is clearer.

How you order information depends on the aim of the report. If management want your recommendations for changing production schedules (because you are the expert) you might start with a summary of these and back them up with reasons or evidence for making them. You could split this into a general discussion about how you reached your recommendations and a further section (perhaps an appendix) giving the detailed evidence, sometimes largely in the form of tables and graphs.

Leave out irrelevant information. If the data were recorded by Mr Smith prior to 1987, but since then they have been recorded by Mr Jones there is no need to say so unless it is relevant (which might be the case if Mr Jones has a reputation for making mistakes).

Some reports do not involve recommendations for immediate action, but may be only for the long term record. For example, tests might be carried out on a raw material from a new source, but there is no intention of bringing it into immediate use. In that case it might be best first to summarize findings on the characteristics of the material and to follow this summary with any technical comments needed about the method of testing, or problems that arose in carrying out the tests.

8.4.2 Writing for technical and trade journals

This is a stylistic half-way-house between writing for professional and learned journals and writing for the popular scientific and technical press. I considered the former in section 8.3, and the latter are covered in sections 9.1 and 9.2. Some of your peers may read your work in technical or trade journals, but the target is essentially other technologists, scientists, or perhaps production and marketing managers, who will have some understanding of, but often not be experts in, your field.

Some of the journals I have in mind are produced and distributed by commercial publishers and others by professional associations or societies, but they report news and general developments in a discipline or groups of disciplines, rather than publishing research papers.

Write clearly and convincingly. Most of these journals have a house-style; some require a rather formal presentation, for others an almost chatty approach may be more acceptable. Articles may be sectionalized using catchy sub-headings; news value and topicality may be important. If the journal is published by a professional body there may even be a 'political' element in the sense that items that enhance the professional status of that body, or bring kudos to it, may be more acceptable than articles that are adversely critical. This does not mean that valid criticisms will be disallowed, but contentious, politically harmful or unsupported criticism or hunches may not appeal. As for all journalistic, or near-journalistic, writing study the market; look at recent issues of publications you are considering as possible outlets. Read any instructions given by their editors; if none are given, try to follow the layout of current articles; e.g., if they start with an abstract or summary include one. If they feature tables and diagrams or graphs use relevant ones. If none seem relevant, and the journal usually features these, it could mean your style or subject is not appropriate for that particular journal.

Check carefully the lengths of contributions. Some journals allow a wide range; others restrict all items to one, or at most two, pages (perhaps between about 600 and 2000 words, the precise count depending to some extent on the number and size of diagrams and tables as well as journal page size). Avoid, unless they are essential, complex graphs and tables; these do not appeal to those who are not specialists in the subject matter.

8.5 ADVERTISING COPY

Advertising copy is usually written by skilled publicity writers working for advertising agencies or public relations specialists. For advertising with a technical bias, these writers will probably be given the necessary technical detail and may send their final copy to their client's technical experts for checking. However, occasionally a technical expert may be asked to produce advertising copy. This calls for a mix of journalistic skill with technical know-how. Although the evidence sometimes looks to contradict this, there are ethical standards that are widely observed in advertising and that have legal backing in many countries. In particular, false or misleading claims are not allowed. You must not claim:

Widget A will decrease your running costs by at least 10%

unless that statement can be substantiated. You will get away with

Widget A will decrease your running costs by up to 10%

if you can show one example, however obscure, where it does so. Also, a blatant, but quite outrageous statement, is usually accepted as ethical if it is such that no reasonable person would take it seriously. A well-known example is the claim that a certain beer *reaches the parts that other beers cannot reach.*

It is also acceptable to list the good features of *Widget A* (low initial cost, on-site maintenance service, enhanced rate of output, etc.), without referring to its weaknesses or less attractive features (high cost of replacement parts, need for regular cleaning or frequent adjustment of operating controls, that it may produce an unpleasant smell if used for long periods in a confined space, etc.). However there may be a legal requirement to attach a 'health warning' to some products, e.g.

Ventilation facilities must comply with relevant US standards.

It was once regarded as unethical to knock competitors' products. This is now done widely, although still regarded by many advertisers as unwise; it is a policy that may rebound; it has the best chance of success if the knocking is witty and not too ill-natured.

If you have to write advertising copy the difficult part is catching the attention of potential buyers. This is not the same as making your advertisement attractive to as many readers as possible; not all readers will be potential buyers. Catchphrases, puns, logos are all widely used eye-catching devices, but some link (often subtle but not too obscure) with the product being advertised is important.

Enjoy our nice hot summer

might be a good advertising line for a pocket calculator. Be careful though not to infringe registered trademarks; if you are marketing a new word-processing package you will be in trouble if you claim it is the only program that *guarantees wordperfect output* and in even bigger trouble if you write *WordPerfect*.

If advertising technical products, or consumer equipment, you may have to include specifications. These are often best given at the end or in a separate panel (often in smaller print). Use of appropriate and attractive typefaces is important in advertising.

On the whole, advertising is best left to the professionals, but if you have to prepare technical advertising copy, study advertisements for similar products carefully; but do not copy them too closely. If you do you may run into copyright problems (Appendix A), and more importantly, potential buyers may confuse your product with the one that has already been described in similar terms.

Many sales brochures are, in essence, advertising — which should be informative. You will probably be employed by the marketing department in your organization if you have to write these; if so, you are likely to have had training in marketing, as well as having the relevant technical know-how.

8.6 WRITING BOOK REVIEWS

Your recompense for reviewing a scientific or technical book is usually an invitation to retain the review copy. If the book is one you want, that is usually acceptable. A book reviewer should aim to be fair to the author, the publisher and to potential readers of the book. I have read reviews that are little more than paraphrases of the blurb on the book's dust-jacket. Potential buyers of a book can read that blurb themselves, and when doing so will remember that its prime aim is to encourage them to buy the book, and assess it accordingly. It may not be clear to them if they read the same material in a book review that the jacket blurb is the source.

A reviewer owes readers more than a rehash of publisher's blurb, or just a list of chapter headings, although for some books it may be appropriate to include chapter headings as *part* of a review. These headings may be relevant if they are informative and give a good idea of the topics covered, or the author's approach to the subject.

Readers of a review look for guidance about who might enjoy, or benefit from, reading the book (e.g. undergraduates, postgraduates, research workers, those requiring an introduction to practical applications, etc.), and for help in deciding whether they should buy their own copy, recommend it for their library, or borrow a copy from some other library. If a target readership is indicated by the author, this should be mentioned and you should say whether the work seems likely to appeal to that target. When reviewing a book for a specialist journal you should also give some indication of how useful readers of that journal are likely to find the book; e.g. is there too much emphasis on theory for readers of some 'applied' journal for whom you are reviewing the book.

Many reviewers find it hard to resist a temptation to point out every inadequacy in a book. It is right to mention any serious deficiency, as it also is to highlight any outstanding feature. Try not to be destructive or overtly negative. For example, if you think an author has ignored important work on his subject do not just say that 'recent work is not adequately covered' but, if possible, be more specific, e.g. state that

> *Work by Smith and Jones (1991, 1993) is not mentioned. This omission calls into question some of the recommendations made by the author in Chapter 9.*

In that case you should give, at the end of your review, details of the references to Smith and Jones. This more positive approach not only alerts readers to a possible weakness in the book, but also tells both them, and the author of the book, where to look for corrective material. In addition, the remark may be helpful to the author if a revised edition is planned.

Even if you are very disappointed by a book you are reviewing and feel you cannot recommend it, still try to be constructive in your analysis of its faults, and if you see potential for improvement in future editions indicate how this might be achieved. I recently reviewed a book that was difficult to read and would certainly be misleading to students because of poor ordering of material (several concepts were used before they were defined), inconsistencies in notation, references being incorrectly cited and serious misprints or omissions in mathematical equations. In my review I gave instances of such faults and suggested that both author and publisher should accept some responsibility for the shortcomings, many of which could have been eliminated by better copy-editing and by more careful proof-reading. I genuinely hoped my remarks would help both the publisher and the author to do better in future, for clearly the author had some good ideas.

9

Popular science and technology

9.1 SCIENTIFIC AND TECHNICAL MAGAZINES FOR THE GENERAL READER

I look now at writing for well-known journals like *the New Scientist* and *Scientific American*; also for technical magazines catering for business and trade users, hobbyists and DIY enthusiasts in areas like computing, motoring, flying, photography and home improvements. These publications have a stronger journalistic flavour, and often a larger circulation, than the professional technical and trade journals that I covered in section 8.4.2. Much of the material for the wider-circulation periodicals is supplied by staff journalists or other writers known to the editor, but many have openings for topical items from freelance contributors.

Timing is often critical. Magazines like *Personal Computer World* review new software packages before or coincidental with their launch. Many such journals go to press a month or more before they appear in shops and on bookstalls. This means that software reviews are only possible if the reviewers have pre-release versions of the software. Unless there are major technical problems with a new package, the software houses are keen to make pre-releases — often called β-test copies — available to the leading journals. A magazine passes these to staff writers or to freelances known to the editor for an ability to test, assess and write a review. Only if you are well versed in testing computer software and have some journalistic skills, are you likely to get this work. Readers will want to know:

- what the new package does (and, if it is an updated version of an established one, what is new);
- how it compares with competitors — in content and price;
- how easy it is to use, i.e. whether it is user-friendly;
- are there any serious bugs?

Bear these points in mind if you review software (and specialist software is now sometimes reviewed also in scientific, technical and trade journals of the type I wrote about in sections 8.3 and 8.4). A software, or for that matter a hardware, review will almost certainly have to fall within specified length limits — e.g., between 500 and 600 words, or not more than 1100 words. You will be in trouble if you stray outside the given limits; a journal must keep to these because not only is space precious, but even more than in books, attractive layout is a sales stimulant, and lengths of items influence layout. Positioning of advertising also has to be taken into account. It is no help to the editor if your article requires one column and three lines or spills by half a line onto a new page, or fills a column but leaves no room for a heading! At best sub-editing will then be needed.

The above example is about contributing at the specialized end of the market. Some general science news magazines — the *New Scientist* is one — provide, on request, a guide to house-style. This covers points relevant to contributing to any journal (and to most writing) such as double-spacing your manuscript. It also indicates other requirements, some of which may or may not apply to other journals. In the case of the *New Scientist* these include:

- give an approximate word count;
- left justify only (e.g. leave ragged right margins);
- feature articles should be between 2000 and 4000 words;
- include a brief biographical note of about 20 words;
- no footnotes;
- avoid references — cite other work only in the text;
- at first use, give people's first names and indicate where they work (e.g. Daphne Smith and Robert Jones at the University of Boncaster . . .), but use surnames only at subsequent mentions;
- send suggestions and roughs only for diagrams; these will be redrawn and labelled to suit the journal's style;
- use SI units unless the original measurements were not SI or an industrial standard is used (e.g. 3.5-in. discs).

Helpfully, as well as listing these and one or two other specific requirements, the *New Scientist* also gives advice relevant to most writing at this level, including the following.

- Write for someone intelligent whose interests lie outside your field (this effectively defines a target readership).
- The introductory paragraph should 'grab' the reader's attention and give a hint of what is to come.

- Avoid jargon. If a technical word is needed, explain it first.
- Avoid passive verbs.
- Avoid ugly noun stacks, e.g. replace *the Westminster government-imposed 40-day fishing ban starts now* by *the ban on fishing for 40-days imposed by the Westminster government starts now.*
- Avoid clichés and tautologies such as *total monopoly, absolute finality.*
- Translate names of foreign institutes, etc. into English.
- Do not imply all scientists (or all readers) are male.

With the exception that in more technical popular journals (e.g. those on computer science, aviation, home improvements, etc.) relevant jargon may be used more freely, you are unlikely to go seriously astray if you adhere to the *New Scientist*'s advice when you write for the popular scientific and technical press. Some of this guidance echoes points I made about microstyle in Chapter 3.

As for any submission to a publisher, do not forget to put your name and address, phone and fax numbers, and e-mail address (if you have one) on the title page. It is a good idea to add your name and address at the foot of the final manuscript page also. Do not staple pages, but fix with a paper clip after checking carefully that all pages are numbered and in the right order. This applies also to the types of submissions I discussed in sections 8.3, 8.4 and 8.6.

The advice that the introductory paragraph should 'grab' readers' attention is important in popular writing. If you are writing about people's conception of relative risks it is not very inspiring if you start:

A number of writers have given examples where people's conception of the levels of risk associated with certain activities do not accord with statistical evidence on the relative frequencies of occurrences.

You will encourage more readers if you write instead:

Are murders as common as deaths from strokes (cerebral haemorrhages)? Many people answer 'yes'. They are surprised to learn that even in countries with a high murder rate, deaths from strokes are up to ten times more common.

Media coverage fuels the popular misconception; a murder is sure of a mention in the local newspaper; those dying from a s.. oke will only make the news pages if they are prominent citizens — even then the cause of death is often not specified.

9.2 WRITING FOR NEWSPAPERS AND GENERAL PERIODICALS

Increasingly, as radio and TV dominate topical news coverage, serious daily newspapers are switching from on-the-spot news to in-depth analysis of current issues; this includes developments in science and technology. These topics do not get as much cover as politics, international affairs, the arts, or the misdemeanours of TV personalities, major or minor royals, cabinet ministers or presidential advisers. However, many top national dailies devote several pages once or more each week to topics like developments in medicine, information technology and computing, biotechnology, communications, the construction industry.

There is also space devoted to social issues including psychological explanations for changes in attitudes towards religious and moral issues, as well as to attitudes to crime, the welfare state, education, etc. While such issues are 'popular' at the time I write, the emphasis is likely to change and it may be quite different three years hence. When writing for the daily press be sure you are matching an editor's current requirements. You may not like an editor's judgement, but editors who consistently misjudge what their readers want don't stay long in the job.

Some newspapers have science or technical correspondents who write on current developments and about innovations. Many of these seek, from time to time, advice from experts on special topics or scientific developments or on technical matters, particularly in controversial areas like the safety of nuclear power plants or the dangers from acid rain or the threat of global warming. A newspaper's science correspondent may sometimes hand over the writing of an article to an expert — particularly if that expert is a household, or at any rate reasonably-well-known, name. More commonly, a science correspondent (or perhaps even a reporter who has no special credentials in science) will be looking for information to be moulded in his or her own way into an article; you may or may not be quoted as the source. Experts are sometimes asked for a direct quote on a controversial issue. In any of these situations if your views are sought you may, and should, charge a fee for professional services, but make this clear before giving an opinion or advice, because the newspaper's representative, whether science correspondent or reporter, has a right at that stage to withdraw the request.

If you are to be quoted, directly or indirectly, as a source of information, reserve the right to see the article prior to publication and to correct any misquotations. Unfortunately a few arrogant editors (the 'editor's decision is final' syndrome) sometimes alter copy at a late stage without consulting anybody — even the journalist who wrote it. In practice there is damn-all you can do about this, other than try to get the editor to publish a correction if you have been seriously misquoted; you may wisely decide never to assist that publication again. If the misquote is such that it might harm your reputation, you could consider suing the paper for libel, or infringement of your moral rights, but that may cost you money, especially if your action fails.

If a newspaper or magazine does not have its own science or technical correspondent there is a better chance that its editor may either accept an unsolicited freelance submission on a newsworthy scientific or technical subject. Some editors may even approach an expert asking for an article on a topical matter. If you are approached, seek clear guidance on the length and type of article required. Enquire whether the paper wants it slanted in any particular way, e.g. to highlight work done in your own laboratory or to emphasize the importance of international collaboration. Check when the article is required — newspapers work to tight deadlines — and what payment is being offered. For a top national UK daily or Sunday newspaper payment between £200 and £400 per thousand words for a solicited article is reasonable; perhaps half these amounts for a major provincial paper, but you may have to screw your local paper for what you can get. In the USA rates are often higher than the dollar equivalents.

If you are a freelance with an idea for a newspaper or magazine article follow the general advice on writing for newspapers and magazines given in the latest edition of *The Writers' & Artists' Yearbook* (FRII), *The Writer's Handbook* (the UK or USA publication of that name as appropriate) (FRII) or *The Writers' Market* (FRII). Because you are writing in a specialist field a preliminary letter is often advisable — but skip this step (or replace it by a telephone call) if the material is highly topical or effectively 'news'. In a preliminary letter outline in 100 words or less the theme of your proposed work and ask whether the editor is interested, and if so, on what terms he or she would be prepared to consider an article. For freelance writing you can cover a wider range of topics than simply new scientific and technological developments. Articles to mark the centenary, bicentenary, tercentenary of the birth or death of a prominent scientist, or of some important discovery or invention — particularly if there is a local or national connection — are often acceptable. For material of this sort, submit your proposal preferably six, and certainly not less than three months, before the event, because magazines and magazine sections of newspapers (the likely homes for such pieces) are often planned several months in advance, and may even go to the printer a month or more before the publication date.

Because of emphasis on topicality, writing for newspapers and magazines is usually a more hurried process than writing a book. Contractual conditions may be vague, but try and get firm agreement about what rights you are selling and what the payment arrangements are. Newspapers are usually content with the right of first publication in a particular country (in the UK known as *first British serial rights*). This leaves you free to sell the material in overseas markets. Resist requests for sale of complete copyright; this may not seem to matter if the material has a local flavour or is topical and unlikely to be saleable elsewhere, but it can lead to difficulties if you want to incorporate the same material in another publication (e.g.a book) at a later date.

All of us at some time have read an item in a newspaper or magazine on a familiar topic and been irritated by, often minor, inaccuracies. When reporting topical or ephemeral material this is almost inevitable. Try to avoid it when writing about science or technology; not only because these are subjects that call for exact or accurate treatment, but because some of your readers, perhaps only a small minority, will be experts and ready to pick up weaknesses. If you are quoting statistics be careful to base them on current information and data, and be sure that they are relevant. There is a fine line between omitting boring and unessential detail (which should be left out in most popular articles) and giving a misleading impression by withholding relevant information (which you should not do). Be careful, too, about terminology; I once incurred the wrath of a manufacturer when I used a trade name as though it were the correct word to describe a common item. For instance, both *Thermos* and *Hoover* are trade names and should not be used as synonyms for *vacuum flask* and *vacuum cleaner*. *Kleenex*, *Frigidaire*, *WordPerfect*, *Westinghouse*, *MS-DOS* and *IBM* are other trade names.

Magazine and newspaper publishers are more erratic than book publishers in arrangements for paying their contributors. My happiest experience was when I sent an unsolicited contribution to an aviation journal and received payment within a week of submission, together with an apologetic note from the editor saying that owing to a backlog of material it would be a year before the article appeared. A few journals pay when they accept an item (and acceptance may come a long time after submission if that item is not immediately topical), but most pay either on publication or sometimes at the end of the month following publication. Regrettably, it is not uncommon for authors to have to press an editor for payment because this has not materialized at the promised time.

9.3 RADIO AND TV

This is a difficult subject to cover in a book like this because requirements of broadcasting organizations depend strongly on the country in which one works. What I say here applies primarily to the British market, although some of my remarks apply broadly to countries such as Australia that have similar broadcasting set-ups. The US situation is not the same, because both public service and commerical broadcasting are organized differently. In Canada, the situation is intermediate between that in the UK and USA. Nevertheless, many of my comments on the situation in the UK apply with appropriate local modifications in other countries.

9.3.1 Radio

A current trend is to divide sound broadcasting into talk and into music channels. Clearly the former, especially if there is an emphasis on news, current affairs and intellectual and educational matters, is of more interest to writers about science and technology; although a few years ago talks on science or the arts were common in concert intervals on the BBC's classical music channel, Radio 3. The BBC are the main radio presenters in the UK of science, medicine, engineering, technology or sociology based programmes, mostly on Radio 4 or on regional networks. However, at the time of writing there is a proposal to introduce a news, current affairs and sports network with a stated intention to devote more time to topical scientific items.

The present networks often feature reports on new developments and also interviews — mostly conducted by professional broadcasters — with workers in their laboratories or, particularly in the case of natural history programmes, in the field. There has been a drift away from scripted talks; however, preparing a report on a new development, or even getting material together for an interview, has aspects not unlike writing a script or a magazine article — except that it is, if anything, more demanding — especially if you are going to broadcast 'live'. In a taped interview, or when recording a feature to be broadcast later, editing can save embarrassments like accidental spoonerisms (converting a *thick brush* to a *brick thrush,* or worse, a *hard chore* to a *charred whore*), 'drying up' in front of a microphone, and other crises familiar enough to a professional broadcaster, and gaping traps for a novice. So prepare your work well and rehearse it as much as possible.

Producers, many now freelances, selling productions to the broadcasting organizations, often seek experts for specialist programmes, but if you feel you could contribute something to an established series check the name of the producer — usually given at the end of each programme — and put your ideas to that person. A BBC regional producer recently circulated all university science departments in his region seeking ideas for programmes in a new series and intimations of interest from would-be presenters.

The demand for nature and wildlife programmes seems almost unlimited. These usually involve going out with a tape recorder (and high-standard recording equipment is essential for the sound quality required for broadcasting). If you have interesting wildlife in your garden you might be able to record their mating calls, or the sound of their battle-cry, interspersed with an appropriate commentary. You are more likely, however, to find a role in a nature programme if you are an expert on the otter, toad, adder or whatever, and collaborate with a producer and presenter to compile a recorded programme at some suitable location such as the creature's natural habitat. Because birds and animals are often unwilling or uncooperative participants this type of programme is seldom broadcast live. Many hours may be spent recording material for a 20-minute programme, so there will be considerable editing.

Medicine, science and technology are moderately common ingredients of the 'chat interview' feature where a well-known presenter (often pretending to be a lot dumber than he or she really is) interviews an expert 'invited guest'. This is the nearest radio now gets to the old-fashioned scripted talk. It has evolved largely because of evidence that many listeners have a limited attention span and a dialogue is more likely to hold their interest than is a monologue; so the former is a preferred way of conveying information.

The dialogue consists basically of the presenter asking questions, perhaps laced with anecdotal material, and the contributor providing answers. If the broadcast is 'live' and the subject scientific or technical, there will almost certainly have been behind-the-scenes preparation. A 7–15 minute interview will typically involve 'answering' some 5 to 10 questions from an interviewer who will have fore-knowledge of your likely answers. This lets the interviewer add an element of continuity by ensuring that each question after the first links to what has just been said, perhaps asking the interviewee to elaborate upon a point that has just been made, or urging a shift to a new theme. A common approach is for the contributor to list (in note form may suffice) ten key points that he or she will elaborate upon in the interview; these notes, perhaps with some modification by a producer or editor, are made available to the presenter, who frames questions accordingly. In practice, not all the pre-prepared material is used in most interviews, for producers have to work to strict time schedules and they always plan so that there is something in reserve in case the unexpected happens.

From this brief description you may think that the whole set-up for interviews is casual and that there is not much planning. This is not so. Even if the material is only in note form, the notes must be sufficiently detailed and arranged in a logical order to ensure a smooth flow of material. While an advantage of unscripted material is that the odd hesitation or slip gives an impression of spontaneity, unless there is careful preparation this may degenerate into chaos, e.g. the questions being asked in wrong order, or answers not corresponding to the question asked. Little slips or hesitations like:

> *He first noticed this, I think, in 1925,. . . er, no it must have been 1926, because he spent all of 1925 in France*

go down well with listeners, giving a hint of human fallibility, forgivable because the person obviously knows their material. But if you are being interviewed about the life of some famous scientist and the interviewer asks you to tell us something about his childhood it will irritate listeners if you say

> *Well, that really wasn't very interesting, but after he married he travelled a lot on the Continent.*

The listener will, probably rightly, conclude that you are dodging the issue because you are ignorant about his childhood, or because you have not prepared your material properly. If a question does creep in that completely stumps you, be honest; if you know nothing about your subject's childhood, but are asked about it, you may do better to say something like:

> *I am afraid you've caught me there; I've never come across any accounts of his childhood. It would be nice to have some information, but his biographers seem to have been rather coy on the subject. Perhaps some of our listeners might be able to help.*

Of course, you must avoid making any such assertion or a call for help if you know the information is readily available, but you have simply not bothered to read it. Many listeners will detect your laziness, especially if you are on a programme with a large audience. Radio audiences sometimes number hundreds of thousands; that is many times the numbers of readers for a reasonably popular book, or for a magazine article on science or technology.

Schools broadcasts are another field for the freelance, but, like writing school textbooks, a highly professional one. Most presenters are experienced teachers and organizations such as the BBC provide guidelines on preparation of schools-broadcast material, both for radio and for TV.

9.3.2 Television

Compared to radio, TV is a big money spinner if you can break in. Both public service TV (which is in Britain the BBC) and commercial broadcasters in most countries feature natural history, wildlife and some other general science programmes. These require either continuity scripts, or some may be based on books.

I mentioned in section 4.3.1 that a 1993 *Equinox* programme was based on Roger Penrose's *The Emperor's New Mind* (FRIII). Even more than in radio, the new breed of independent production companies are ever on the lookout for ideas for TV programmes either for established series like *Equinox* or *Horizon*, for school or other educational programmes, or for new one-off features. Scientists may provide material either in the form of scripts (television scripts have a special format that you will have to learn if you hope to become a script writer), or they may be interviewed in front of video cameras, or provide basic material which producers will arrange to have scripted; the scientists will either present this themselves or it may be spoken by professional actors against some relevant background filmed material. Sometimes, if you have provided material in the form of film that requires a commentary you may be asked to do that commentary yourself. Scripting this is no easy task, as it must be carefully timed to the film; even after it is scripted getting the timing synchronized with the visual is not easy. Producers and their assistants can be helpful in guiding people in using several studio tricks to help timing.

For major feature programmes producers look for a 'tag' on which to hang a single production or a series. Often this will be something topical, perhaps a recent spectacular earthquake or a political row over proposed environmental legislation. They will look, too, for a controversial angle to make people not only 'want to watch' but also to 'keep watching'. The producer will also want to see varied and simple explanations and examples that will be familiar to and appeal to that mythical character 'the man in the street', e.g. a programme about risk might introduce examples from gambling, insurance, new medical treatments and travel.

9.3.3 Broadcasting fees

Fees vary greatly between different countries and broadcasting organizations. In 1993 the BBC rate for the now rather rare scripted and read radio 'talks' ranged from £10 to £17.50 per minute — the higher rate being applicable for professional broadcasters. The rate for scripts only, ranged from £8 to £13.50 per minute. These figures are for network broadcasts; lower rates are paid for material broadcast in one region only. Rates for unscripted short talks of about 5 minutes' duration ranged from £29 to £46.

Rates for presentation of features and documentaries depend on how much the individual does; someone, usually a professional broadcaster, who does the complete job — presentation, linking, setting up and conducting interviews, editing and research for a network radio broadcast — expects to receive about £23.50 per minute of broadcast time and about two-thirds of that rate for a broadcast in one region only.

Television rates are usually appreciably higher and negotiable within certain limits. The UK legal requirement that the BBC and other broadcasters now purchase a proportion of their output from independent producers has opened up opportunities for scientific and technical writers to negotiate their own fees with these producers (and most are tough negotiators; they need to be to survive in a highly competitive market). Any specialist writer in the UK who gets deeply involved in writing for radio or TV should join a professional body such as the Society of Authors' Broadcasting Writers' Group or the other professional writers' body, the Writers' Guild, which takes a special interest in writing for radio, TV and films. In other countries writers for radio and TV would be well-advised to enquire whether there are corresponding bodies in their country and to consider joining any such body. Experienced writers in these fields are likely to be the best sources of information about the efficiency of any such bodies. I say more about writers' professional organizations in section 10.4

There are a number of books about writing for radio and TV, though I know of none that deal specifically with writing for scientific or technical programmes. The BBC publishes and frequently updates Norman Longmate's *Writing for the BBC* (FRIII) which is a guide to writers on a range of possible markets for their work in both radio and TV. There is a wealth of material about various aspects of broadcasting in Rosemary Horstmann's *Writing for Radio* (FRIII), although it contains little specifically directed to those writing about science and technology.

10

A last round-up

10.1 WORKING WITH YOUR COPY-EDITOR

I cover miscellaneous topics in this chapter. The first is copy-editing. I outlined the copy-editor's role in section 5.1, mentioning Judith Butcher's *Copy-Editing* (FRI) as the standard British guide to their work. Many publishers provide copy-editors and authors with in-house guides that pay special attention to house-style. It is not a formal part of an author's remit to make life easier for copy-editors, but taking simple steps that do often pays a dividend. It saves hassle at later stages of, and may also speed, the publishing process — or at least reduce delays in publication. Most of these steps are common-sense ones. Be careful with spelling, punctuation, grammar, layout and other essentials of style, and comply with instructions or advice for authors issued by your publisher or that are agreed with your commissioning editor. Not only does this speed publication, but an inevitable spin-off is that pressure on copy-editors is reduced.

Here are some ways to save your copy-editor — and often yourself — unnecessary work:

↕   If your book is to appear as one of a series, and a style has been set for the series, comply with that style in your manuscript: e.g., if figures are numbered consecutively throughout books in the series as Fig. 1, Fig. 2, . . . rather than numbering them afresh within each chapter, i.e. as Fig. 1.1, Fig 1.2, Fig 2.1. . . , adhere to the former system in your book — even if you don't like it!

↕   Check for consistency in punctuation and spelling (hyphens are often a problem). Run every text file through your word-processor's spelling checker; this won't pick up everything but it will find those inevitable typos like *thier, understnadable, overrrated* that we all produce from time to time.

☞ Check your numbering system for sections, tables, figures (plates if any), equations and formulae. If you have equations numbered (5.4), and (5.6) to (5.10), but no equation (5.5) and you refer often to your numbered equations a lot of careful checking will be needed to put this right.

☞ Make sure that all citations in the text are listed in the *References* or *Further reading* lists, and that the *References* do not include items that are not cited. A *Bibliography* or a *Further reading* list may include items not cited in the text.

☞ If you are uncertain about some matters yourself (e.g. whether something should be put in bold or italic; whether you should display the formula $I=Prt/100$ on a separate line or include it in a text line as I have just done), draw the attention of the copy-editor to your concern. I suggest below a way to do this.

It is part of a copy-editor's duty to check such matters, but by dealing with them as you write, or before submitting your final manuscript, you are providing a double check. Even the best copy-editors are not infallible, so this double check helps reduce errors to a minimum. This is particularly important for specialist material. Most publishers try to use a copy-editor familiar with the subject, but this can only be done on a broad basis. For example, if you are writing a mathematical monograph, a copy-editor with a mathematical background will be able to check if you have matching brackets and parentheses in a formula, but if you write down an expansion for a Bessel function of the first kind when you meant to write down the expansion for a Bessel function of the second kind it is unlikely that a copy-editor will spot your mistake. That is not part of their job. Similarly, if you are a geologist and state that *basalt* is the dominant rock-type in a remote location when you meant to write *quartz* this is unlikely to be queried by a copy-editor.

I mentioned in section 5.1 that after checking your manuscript (and this applies even to manuscripts submitted as CRC), the copy-editor usually — and certainly should — come back to you with a set of queries. Each query should give the folio (page) number, the line on that folio giving rise to the query, the point being queried, a proposed action and ask the author either to approve that action or suggest an alternative. If you want to put any queries to the copy editor when you submit your manuscript it is a good idea to use a similar format. Here is an example. At various places in this book I have highlighted several features by indented lists with each point highlighted by what is known to printers as a *bullet*. Several different bullets are available. I used the first bullet on folio 1 in line 12 (counting each line, including headings, containing text). There I used a round bullet, ●. There are a number of symbols commonly used as bullets. These include:

In the list on pp.162–3 I used the bullet, ☞. I did this for illustrative purposes. I want to know if the bullet I have used generally, ●, is appropriate, and I would formulate my query this way:

> Folio 1, line 12 (and elsewhere throughout). Is the ● bullet OK, or would ☞ be better?

This query refers to bullets throughout the manuscript and adding *and elsewhere throughout* after designating the first occurrence makes this clear; there is no need to list the many other folios where I have used bullets. This example is relevant for CRC or material submitted on disc with typesetting codes inserted. If the manuscript has yet to be typeset, the publisher's in-house design team will usually decide the appropriate bullet to use.

Typical matters you might query when submitting a final manuscript include the following.

- Is there a preferred spelling for a particular word (e.g. *sulphate* or *sulfate, fertiliser* or *fertilizer*)?
- Ask in a specific instance for a decision as to whether one should use two words, a hyphen, or one word (e.g. *proof reader*, *proof-reader* or *proofreader*);
- Is *Fig.* or *Figure* preferred (e.g. should I refer to *Fig. 3.2* or to *Figure 3.2*)?
- Should I spell out a particular abbreviation (e.g. *VIP*) the first time it is used?
- Should popular names of plant species be given in brackets after the botanical names?
- Would the proof of a theorem, or perhaps the description of an experiment or list of properties of certain chemicals, be better moved to an appendix?
- Should certain definitions, formulae, technical terms, etc., be (a) in bold, or (b) in italic, or (c) displayed?
- Is the format and numbering system used for section headings, sub-headings, tables and figures, equations, etc. OK?

It also helps the copy-editor if you indicate when submitting a manuscript if there are special reasons for adhering to any convention you have used, especially if you know this is contrary to accepted house-style. For example, if, when writing about aviation you have consistently used the units commonly used by pilots and air-traffic controllers (section 3.4.2) rather than, say, SI units, make this clear and say also that you have done so because the majority of your readers will understand these more readily. This could save a copy-editor many hours making unnecessary conversions to SI units. Similarly, engineers sometimes use *j*, rather than the more familiar *i*, for the square root of - 1 (to avoid confusion with *i* as a symbol for electric current). If you adopt this convention point this out when submitting your manuscript. If it is essential that a figure or table should appear at a *precise* point in the text, or be of an exact size (e.g. to provide a comparison with some known object) draw attention to this also; the copy-editor knows how to mark the manuscript so that the typesetter will deal with this correctly.

If you have used special symbols (e.g. Greek letters, mathematical symbols, Hebrew or Chinese characters) and have written these by hand, or you are not happy about the way your computer printer reproduces these, give a list with the symbol as written or displayed, followed by a definition in words, e.g.

μ	Greek lower case mu
$\approx$	approximately equal to
к	Greek lower case kappa
Д	Cyrillic capital D

When your copy editor sends you a list of queries, these will often be on a form that includes a space for your reply. If that space is insufficient, put your response on a separate sheet, cross-referencing it to the query (which is usually numbered). A typical query sheet might have entries like those in Figure 10.1 The folio numbers in that figure do not refer to this book. The notation *5 up* in the column headed *Line* is used to indicate the fifth line from the bottom of the folio.

Ideally your responses should, like the queries, be clear, brief and unambiguous. You may have no strong view on some proposed change; if so it is sensible to accept it. At other times you may prefer not to make a change, yet can see logic in what you are being asked to do. It is appropriate then to say you prefer your original but that you will accept the change if it is thought necessary or desirable. If you consider a proposed change is wrong you should — and are entitled to — say so (see the comments on *authors' moral rights* in section A1.3).

Typically the responses to the queries in Figure 10.1 might be:

Query 1.	Fig. 2.1 is correct.
Query 2.	Yes.
Query 3.	See sheet at end.
Query 4.	textbook.

COPY-EDITOR'S QUERIES TO AUTHOR

Query no.	Folio	Line	Query	Response
1	7	12	'Fig. 1.2' <u>not</u> 'Fig. 2.1'?	
2	11	5up	Display formula on separate line?	
3	17	14	Jones (1953)? Jones (1952) in references. Which?	
4	22	9up	Here and throughout *textbook* rather than *text-book*?	

Figure 10.1 An example of a copy-editor's query sheet.

Your answer to query 3, put on a separate sheet because of its length, might read:

> Query 3. Jones (1953) is correct. Insert
> additional entry in references, folio 217
> after line 15:
>   Jones, W.W. (1953). How red
>   is red? *Journal of Nasty*
>   *Colours*, **17**, 221-29.

Other ways to help your copy-editor include care with detail on matters of house-style. Here are examples of two common conventions included in many house-styles (but these may be overruled by your own publisher's house-style and if so, you should follow the latter).

- In headings of chapters, sections, etc. unless capitals are used throughout, keep capitals to a minimum. e.g. write

2.3.1 Technical considerations

rather than

2.3.1 Technical Considerations

Do not insert a full point at the end of headings.

- It may seem perverse, but the usual practice is to put no punctuation at the end of a table heading, but to put a full point at the end of a figure caption. The justification usually given is that a table heading refers to all that follows (and is therefore somewhat analogous to a section heading) while nothing (directly relevant to the figure) follows a figure caption; it is essentially a complete paragraph.

Be especially careful when checking, that there is a complete match, apart from permissible exceptions of the type indicated in section 2.2.5, between listed references and those cited in the text; also that you are consistent throughout in the way you give references; e.g., do not abbreviate the title of a journal in one place and give it in full in another. Remember that it is not part of a copy-editor's job to check the accuracy of references. THAT IS THE AUTHOR'S RESPONSIBILITY. Be particularly careful to give correct year, volume and page numbers. Mistakes may also occur in spelling authors' names, in the titles of papers and sometimes even in the names of journals. Be especially careful when journals have similar names, e.g. *Biometrics* and *Biometrika* or *The Journal of Ecology* and *The Journal of Applied Ecology*. There are few things more irritating to a reader trying to track down a reference than to find after a half-hour search that it has been wrongly cited.

10.2 MULTI-AUTHORSHIP

10.2.1 Joint-authorship

Books, papers and reports are often the work of more than one author. This may result from collaborative research or a joint development project. The physical division of the writing between authors of the book, paper or whatever, may take various forms. Naming each collaborator as an 'author' in a joint project may be primarily an acknowledgement that each took part in the project. The actual writing may well have been done by some subset, perhaps only one, of the named authors; in this case it is proper, and likely, that all named authors will have been asked to comment on the text at various draft stages. In this situation there is usually little difficulty in producing a coherent manuscript unless unfortunate tensions arise between the active author(s) and other collaborators during the writing process.

A second basis for joint authorship is one where authors with complementary interests agree to collaborate. This situation is common when preparing monographs, textbooks, or perhaps an expository review of some specialist topic or subject area. For example, a biochemist and a pathologist may collaborate in preparing a monograph on some aspect of diagnosis of cell abnormalities that indicates a breakdown in the body's immune system. Here, some sections (probably even complete chapters) will almost certainly be written primarily by one author, and other parts of the book by the other contributor. Because no two writers have the same style, it is almost inevitable that there will be noticeable stylistic differences in the early draft material. Unless the authors are prepared to compromise and attempt to smooth such differences, this will increase the load on the copy-editor, and if the copy-editor slips up, the book will be less readable than it should be.

Most writers have different *microstyles,* but problems of joint-authorship often show in some facets of *macrostyle.* For example, if Author *A* writes the first draft of Chapter 2 and author *B* the first draft of Chapter 3, it may be that in Chapter 2 some concept is discussed very briefly on the assumption that all readers will have the necessary background information to understand what is relevant, whereas in Chapter 3 there is a lengthy discussion of the concept that gives a detailed account of all relevant background information. Clearly in this situation each author is making different assumptions either about the target readership, or if they are agreed on their target readership, about how well informed that readership is likely to be on some matters.

Another difficulty with this type of collaborative writing is overlap. Author *A* might describe certain aspects of cell structure, or of relevant cell deformities, in one chapter, and author *B* present virtually the same information several chapters later, perhaps confusing the issue by using different terminology or a different emphasis. The authors must get their heads together at the draft stage to eliminate such inconsistencies.

Problems of this kind may be minimized by sensible preplanning. Compiling a reasonably detailed synopsis of what each chapter should contain is clearly a good starting point. This forces early collaboration between authors. While obviously there must be a broad agreement at an early stage of a project about who is going to write about what, it may well be helpful to prepare chapter synopses collaboratively even before deciding *precisely* who is going to write any particular part.

Even with clear-cut preliminary plans it is important that each author sees at regular intervals draft material from the other author or authors. All should concentrate, in particular, on doing what is needed to ensure that their work dovetails smoothly with that of the other contributor(s). As well as looking at steps needed to harmonize *macrostyle* features, keep a close eye on *microstyle* as the writing proceeds. A book will jar on readers if one writer uses mainly the passive and the other the active voice. Similarly if one uses *I* and the other *we*, there is further irritation. Some copy-editors are more alert than others at spotting such differences.

If, in a joint-authorship situation of this kind, one author is a more experienced writer than the other, or others, it may be to their mutual benefit if the more experienced one is given the task of 'editing' the completed first draft for consistency of both macro- and micro-styles.

While there are many successful collaborative partnerships in which each author brings complementary information, or each enhances the contribution made by the other, sadly, some joint writing projects also fall by the wayside or finish up as unsatisfactory compromises because of human incompatibilities. Difficulties may range from failure to agree about who is to write what, about how certain aspects of the subject are presented, or there may be timing difficulties — one author failing to keep to agreed schedules. This may not only result in a contractual or publishing schedule not being met, but in this age of rapid development in many subject areas the projected book may be overtaken by events — someone else gets in first with a similar book, or new developments make the project out-dated. This is heartbreaking to members of a partnership who have kept their side of the bargain.

Do not enter a collaborative writing partnership lightly. Ideally, do so only if you are confident you can work in harmony, that neither you nor your partner(s) are over-committing yourselves, and that you are sure you can all keep to any agreed schedule. If, as the project develops you see serious difficulties about collaboration, it may be wise to seek an amicable agreement to terminate the project or to reorganize it, perhaps with changes in the team of writers.

A final point — had I been writing this book a few years ago I would have advised against joint-authorship 'at a distance', i.e. where the collaborators were in different institutions or even on opposite sides of an ocean. Fax and e-mail now enhance the prospects of success for such ventures, although I still doubt whether these new technologies carry all the advantages of regular personal contacts.

Summing up, joint-authorship projects can benefit greatly from frequent critical self-editing by the *participants*.

10.2.2 Contributing a chapter

Scientists and technologists are often invited to contribute a chapter to a book, or they may be required to prepare a conference paper for publication in a volume of *proceedings*. These tasks are not equivalent. Books with contributed chapters are often conceived by publishers themselves, or a proposal for such a book is put to a publisher by experts in some field. There is usually a carefully planned structure from the outset, whereas conference proceedings are more loosely structured and are simply a record of all papers accepted by the organizers for presentation at a particular conference. In extreme cases there may be virtually no restrictions on participants presenting whatever paper they wish at a conference.

I consider first the invited contribution of a chapter, or chapters, on a particular theme. Many books produced in this way are in essence a collection of short monographs, each on a particular aspect of a defined topic. The inspiration is often a realization by the book's sponsors — a publisher or experts in a field — that recent developments create a need for a book to report these, yet these developments are so diverse that no single writer is appropriate. The subject might be *Modern developments in molecular biology* or *Competing theories on the origins of life on earth* or *Recent developments in bridge building* or *Forecasting the spread of AIDS* or *A critical comparison of computer spreadsheet software.* The editor, acting on advice from experts — who often constitute an editorial panel — will usually invite appropriate people to contribute one or more chapters. Your reward for participation may be only enhanced prestige, glory, or smug self-satisfaction at being asked. But if the project is sponsored by, or has been accepted by, a commercial publisher, expect some payment. This may be on a royalty basis, but more often, you will be offered a flat fee. The offer may depend on:

- the likely market for the book;
- the length of your contribution;
- your reputation in the subject area.

There is often a mixture of commercialism and prestige in such projects, so I duck out of giving firm advice about payment and only suggest that you should not be too greedy, but that you should not take on the project if any fee plus the likely prestige value does not appeal to you. You are in a half-way house here between writing a book (for which most authors expect some payment) and writing an expository paper for a learned journal (for which you will often not be paid, although a few prestigious journals do give an honorarium to authors of invited expository contributions).

The editor of the book, whether a member of the publisher's staff, or an outsider who has proposed the book, should give guidance to invited authors on:

- the required length of the contribution;
- the form in which references are to be given;
- any special layout requirements such as structure into sections or restrictions on coverage;
- use of tables, diagrams, photographs and how such material is to be submitted;
- whether CRC or manuscript on disc is acceptable, or is required.

There is often some latitude about the length of contributions; the instruction might be 'approximately 10 000 words', or 'between eight and ten thousands words' or in similar vein. You may be asked to give a word count when you submit.

References are usually required in either the Harvard or the Vancouver systems (section 2.2.5), with a complete list at the chapter end. If the instructions you get are not clear, seek clarification at an early stage; this will save a lot of work for both the editor and yourself. For references, you need to know, for example, whether titles of articles in journals are to be included, whether journal names are to be abbreviated and whether first and last page numbers are to be given, or only first page numbers. You may want also to clarify the reference situation for multi-author papers, e.g. are all authors to be named or only a specified number followed by *et al.* There may be latitude for authors over matters like dividing work into sections, but for consistency between chapters there should be guidance on the numbering system to be used for sections and subsections (and further subdivisions if these are permitted). Sometimes there are specific requirements for layout; e.g. that each contributed chapter is to commence with an abstract, followed by an introduction, followed by a short historical perspective, with the author left free to decide the layout after that point.

Tables and diagrams are often required in specific formats, sometimes prepared as CRC, even when this is not used for the body of the text. There may also be special instructions on how headings and captions are to be displayed. Use of photographs may be restricted to those accepted as essential after discussion between chapter-author and editor.

If manuscript in camera ready copy form is acceptable, or is required, ask for full instructions about layout. Unless the publisher is casual about layout (an attitude that will do nothing to enhance the book) it is essential for all contributors to use the same page format and type fonts and point sizes.

If the author is to supply running headers for each page, instructions should be given about how these are to be presented. For obvious technical reasons with a multi-author work of this kind, the final page numbers will have to be inserted by the publisher when all contributions are collated. To ensure uniform text areas on all pages you may be supplied with paper with blue margins and be told to align type to, and to type only within, these margins. The blue margins disappear in the photo-setting procedure. Where relevant, you should also be given instructions about format requirements for manuscript on disc that is to be used for typesetting without re-keying.

As with all writing, when preparing a chapter for a book, you will save everybody (including yourself!) work by sticking to editorial guidelines and not hesitating to seek clarification of these if you need it.

At the start of this subsection I hinted that there were similarities, but not a true parallel, between writing a contributed chapter to a book and preparing a written version of a paper presented at a conference for publication in a volume of conference 'proceedings'. A main difference is that whereas a multi-author book usually has a clear theme, and authors are generally asked to write on a particular aspect of a topic on which they have expertise, conference papers normally are less restricted in their requirements. Most often they will consist of reports of new work by individuals or by research groups, or else of expositions on recent developments in some subject area. Most editors or publishers of such proceedings seek some uniformity of presentation covering, for example:

- the way references are given;
- whether an abstract or a list of key words is required;
- whether material is to be presented under specific headings such as *Introduction*, *Material and Methods*, *Conclusion*, *Discussion*, *Recommendations*, etc.

Because there are always pressures for quick publication of 'new' research results, CRC is often required. Sometimes there are pre-conference refereeing procedures and the final CRC must be prepared after these. A few years ago incompatibility between word-processing packages or printing equipment and editorial requirements made it difficult for some authors to comply with the latter. The difficulties are decreasing, but manuscript on disc has an appeal in this situation because it gives professional typesetters scope for reducing inconsistencies between material from different contributors. The poor standard of layout in many CRC produced conference proceedings published in the 1970s and 1980s tended to bring CRC into disrepute, but it does speed publication. I say more about this in Appendix B.

10.2.3 Contributing to reference books

Dictionaries and encyclopedias dealing with particular disciplines have been growth industries in recent years. The main publishers of these usually have staff or freelance editors with expertise both in the requirements for reference works, and a broad knowledge of the subject matter covered by the work they are editing. Relevant aspects are reflected in the instructions they give to (usually invited) contributors.

I contributed to a dictionary of mathematics for which there were two named editors and ten named contributors. My brief was to compile entries in statistics. I was given the following.

- Clear instructions about the target readership (if 'readership' is the right word for those who consult dictionaries!).
- A number of 'specimen' entries prepared by the editors on topics in various branches of mathematics to indicate (i) the required level of treatment, (ii) conventions for the choice and display of headwords, (iii) requirements for cross-referencing and (iv) how examples were to be used to illustrate the application of a technique.
- Notes on the way entries were to be prepared, e.g. each entry on a fresh folio with headword in bold, double spacing throughout, and any example to be at the end of an entry. Folios to be arranged (un-numbered) in alphabetic order by headwords.

It almost goes without saying that great care is needed to get dictionary entries accurate and intelligible. I wrote 'almost goes without saying' deliberately, because unfortunately a number of specialist dictionaries do not achieve these aims. Inevitably, in highly specialized material, the odd blunder will slip in. I found cross-referencing was a bug-bear. Almost inevitably, a contributor will need to give cross-references to material that is being prepared by another contributor. It is unlikely you will see this material (at least at the writing stage). It is probably wise to put in a marginal note for the editor, drawing attention to where you are referring to somebody else's contribution just in case, for some reason, that contributor does not supply the expected entry, or perhaps uses a different headword to the one you 'guessed' would be used.

You are more exposed to the foibles of reviewers with reference works than with most other kinds of writing. Reviewers seldom read a whole dictionary or encyclopedia; most will look at the entries for a 'sample' of words — often a sample strongly biased towards topics they know something about. Picking holes in 'authoritative' reference entries is many a reviewer's favourite pastime. It gives them as much fun as finding fault with *how-to-write* books.

10.3 WHEN ENGLISH IS NOT YOUR FIRST LANGUAGE

There are pragmatic reasons why many scientists and technologists write in English when this is not their natural tongue. These include the following.

- Many publications prefer, or even require, contributions to be written in English.
- Scientific work written in English often has a larger potential readership than it would if written in any other language. An exception may occur if the topic is of prime interest to a community where only a minority understand English.

Worldwide, many scientists whose first language is not English speak and write better English than a lot of British, American, Australasian or English-speaking Canadian authors. For them, there is no problem other than those common to all writers. However, there are others who have problems, especially with subtleties of grammar or idiom.

The editor of a journal, or a British or American publisher, who is not sympathetic to such difficulties is not a good editor or publisher. So, if English is not your first language, do your best. Explain this in your submission, even if you are sure it will be obvious to anyone reading your work. It may not be, for as I have already explained, some natural English speakers do not write as well in English as you may do.

The attitude of many journal editors is summed up by Hinkelmann (1993):

> . . . poor English (by non-English speaking authors) in itself is not a reason for
> rejection, but the better the English the easier it is for the reviewers to read and
> evaluate a manuscript, as improper English often leads to misunderstanding.

If your grasp of English is so slight that you cannot put your ideas into words without a struggle your wisest course may be to employ a professional translator, although even a professional may have difficulties if he or she is not an expert in the subject-matter of your work.

If you feel you can make yourself understood in English, even if you are conscious of difficulty with grammar and idiom, you may do best to submit your work in English — perhaps seeking first, the aid of an English-speaking colleague, or a friend, to check and polish your writing. If you have no such help submit your work, in the best form you can manage, to an editor or publisher, explaining that you have difficulty with the language.

It is impossible in a book like this to discuss all potential language problems; especially the use of idiomatic expressions, or how to choose exactly the right word if you have only a bilingual dictionary. However, difficulties resulting from these points can usually be overcome by a sympathetic editor or copy-editor.

A typical sentence that needs some adjustment is:

If item it is dropped damages may be incurred.

Any editor would understand this to mean what is more correctly recast as:

If the item is dropped damage may occur

or, even better,

Dropping the item may damage it.

Difficulties with matching singular nouns to singular verbs, uncertainty about using the active or passive, when to use the definite article (*the*) or an indefinite article (*a, an*), choosing the exact word, are understandable problems for the non-native speaker of any language. British and American editors know that English is the nearest thing to an 'international' scientific language. This makes it one of their duties to smooth out the translating problems of those who can at least express their basic ideas in English.

So take heart if you are writing in English when this is not what is idiomatically referred to in English as 'your mother tongue'. The phrase 'mother tongue' is, I am afraid, what some now regard as politically incorrect language. But I must stop before I confuse you even more.

If you have understood this section, except perhaps the last paragraph, you should be able to express yourself in understandable English.

10.4 WHERE TO LOOK FOR HELP

This section is relevant for all scientific and technical writers. Your main sources of help should be:

- books about writing and publishing; a number are included in the Further reading list that starts on p. 204;
- your publisher's editor, or the relevant journal editor; or if writing a manual or report, whoever commissioned you to do so;
- professional societies or associations for authors;
- your literary agent (if you have one!);

- a colleague or friend who is experienced in the kind of writing you are doing.

Books contain a wealth of information on technical matters to do with writing, but do not always provide an answer to a specific problem. It is then that someone in the publishing world — usually your editor — may be able and willing to help. Sadly, some editors are less willing than others. On the other hand, if you have a literary agent — but I have already mentioned in section 4.3.2 the difficulty in finding one — it is that agent's duty to help his or her client, the author.

In most countries there are professional bodies representing authors. The Society of Authors is the main such society in the UK. As I explained in section 4.3.1, the snag is that you cannot join the Society until you have a firm offer to publish from a publisher. You may need help before you reach that stage, especially if you are having difficulties over a contract. A colleague, or friend, may then be your best source of advice if you are unhappy about an offer from your potential publisher.

I advise any book author to join an appropriate professional writers' association in their own country. For example, the UK Society of Authors (details are given in the *Writers' and Artists' Yearbook*) has a sub-group that deals specifically with matters of interest to scientific and technical writers. There is also a medical writers' group, a broadcasting writers' group and an educational writers' group, all of which may be relevant to some readers of this book. The Society gives advice to individual members on contracts and on tax matters and also provides a legal service to advise on disputes over contracts, copyright etc. As I indicated in section 9.3.3, the Writers' Guild of Great Britain represents many writers for film, radio, television and the theatre in particular. Further information about these two UK organizations may be obtained from:

- The Society of Authors, 84 Drayton Gardens, London, SW1 9SB.
- The Writers' Guild of Great Britain, 430 Edgware Rd, London, W2 1EH.

Other relevant UK organizations include the Association of British Science Writers and the Institute of Scientific and Technical Communicators. Prominent authors' societies in the USA are the American Society of Composers, Authors and Publishers; The Authors League of America, Inc.; The Writers' Guild of America. Australian writers are catered for by The Australian Society of Authors and the Australian Writers' Guild Ltd., whose activities are closely aligned to their UK counterparts, while Canadian writers are covered by the Canadian Authors Association, The Writers' Union of Canada and the Writers Guild of Canada.

10.5 LENDING AND COPYRIGHT

In the UK, Germany, and several other countries authors are entitled to payment based on the frequency of borrowing of their copyright books from certain libraries. The UK scheme is called *Public Lending Rights* (PLR) and it covers all public libraries but not academic libraries. It is basically for UK resident authors and each book must be registered after publication with The Registrar, PLR Office, Bayheath House, Prince Regent Street, Stockton-on-Tees, Cleveland, TS18 1DF.

There are a few restrictions on eligibility but the main rules and a description of the scheme is given in each edition of the *Writers' and Aritsts' Yearbook* (FRII). The exclusion of academic libraries means that authors of scientific and technical books tend to do less well than writers of romance or crime fiction, but it is worth registering, especially if you are writing at the popular end of the spectrum. Sometimes when a book goes out of print you will be surprised at the increase in library demand for it.

If you are not resident in the UK, make enquiries among your fellow authors or from a professional writers' group to find out whether your country has a similar scheme.

Under certain circumstances you may also be eligible for a payment for photocopying or other reproduction of any work for which you hold copyright. Again professional authors' societies can often advise on these matters.

Appendix A

Copyright and other legal matters

A1 COPYRIGHT LAW

A1.1 Copyright and licensing

British copyright law is enshrined in the Copyright, Designs and Patents Act, 1988 and US law in the 1976 Copyright Statute of the United States. Two international copyright conventions — the Berne convention and the Universal copyright convention — provide certain standards of 'international' copyright protection in a number of 'member' states which now include most major markets. The legal implications of copyright laws — British, American and International — are fully explained in the Copyright and libel section of each edition of the *Writers' and Artists' Yearbook* (FRII) and I list two books dealing with copyright under Further Reading (FRIII).

This appendix summarizes a few points that are often relevant to scientific and technical writing. In essence, copyright protects an author's form of expression rather than an idea. In particular, 'literary' and 'artistic' works are protected, where, apart from more obvious components, literary works include tables and computer programs and artistic works cover diagrams and graphs. Under British law copyright in a literary work exists once the work is recorded in any appropriate form; this includes a written manuscript, or where relevant, a sound recording, or on film or on a computer disc. No formal registration procedure is required. Under US law, to establish copyright a notice of copyright must be placed on all publicly distributed copies of a work. A standard format for this book would be:

© 1994 Peter Sprent

although certain modifications are acceptable. There are some other non-compulsory provisions for registering US copyright. Procedures vary in other countries, but tend to follow the less formal British pattern. However, there are odd quirks in some countries, so authors should always seek advice when in doubt.

From 1 July 1995 the European Council of Ministers' *Directive on the Harmonization of Protection of Copyright and Certain Related Rights* is expected to come into force. The full implications, especially with respect to subsidiary rights and use of copyright material in, for example, anthologies are not completely clear at this stage. One important stipulation is that copyright will remain in force (at least in EC countries) for 70 years after the death of an author.

To be eligible for copyright protection a work must show originality. At first sight it may seem, therefore, that a monograph which collates established results in some subject would not qualify. However, most monographs meet the requirement because *originality* applies to the way ideas are expressed rather than to ideas themselves. This means a work need not be innovative, so long as it establishes itself as indicating skill and labour on the part of the author. It does not suffice to make an exact, or even a near copy of the original; indeed to do so without permission would be a breach of another person's, or persons', copyright if the material copied were itself under copyright. In general, copyright subsists for 50 years after the death of the author; or of the last surviving author, in the case of multi-authored material. As mentioned above this is likely to be extended to 70 years in EC countries after 1995. Not surprisingly, an exact or little altered copy of non-copyright material cannot be copyrighted, so you cannot claim copyright if you bring out a book consisting of the writings of Galileo or a facsimile of Newton's *Principia.*

Titles cannot be copyrighted, because they are too short to be deemed literary works. However you may be in legal difficulties if you try to pass off your work as that of another author, e.g., do not call your book *A Short History of Time* and publish it under the pseudonym of Steven Hawkyns.

Initially, copyright resides with the creator(s) of a work, e.g. the author(s) of a book or the person who takes a photograph. An important exception, relevant to some scientific and technical writing, especially to the preparation of instruction manuals and technical documents, is that the copyright of a work produced by an employee as part of his or her employment belongs to the employer unless there is a specific agreement to the contrary.

Authors may assign their copyright to another party, typically a publisher or agent. Many publishers of professional and learned journals ask authors to assign copyright of papers to them. Refuse if you can, because otherwise you lose all rights and all control over the work. In the case of books I warned of pitfalls of assignment in section 4.3.1. Generally, it is better to exploit the financial and other gains from ownership of copyright by licensing specified rights to a third party (usually a publisher, or a broadcasting or film producer); this has been the common practice in the UK for many years and under current law is becoming more common in the USA. Until the present Statute came into force, US copyright laws were a dangerous jungle, especially for writers who were not US citizens or resident in the United States

Specific rights may be licensed to different parties; e.g. hardback rights to one publisher, paperback to another and TV rights to an independent production company. Literary agents often make arrangements like these on behalf of the authors they represent, but in the UK especially, it is common practice for an author with no agent to license (often for a limited period) all rights to a publisher, who may then sell some of these as subsidiary rights under a secondary licence to various third parties. It is for these reasons that it is important when considering a publisher's contract to be clear about

- precisely which rights you are assigning;
- what proportion you are to receive of any revenues the publisher obtains from the sale or exploitation of any subsidiary rights

A1.2 Permissions and infringements

If you reproduce any copyright work without modification, or with only cosmetic changes, without the permission of the copyright holder you are, with certain exceptions indicated below, infringing the rights of that holder. Similarly, anyone who does the same with your work is infringing your copyright. This applies to any form of reproduction including — again with certain exceptions — photocopying and translation. If rights granted to a licensee are infringed, that licensee is also an aggrieved party. Clearly there is a grey area as to what constitutes what I have called *only cosmetic changes,* which, incidentally, is not a term used in either UK or US copyright law. Copying in this context is usually taken to be proved if one can show substantial similarities between the original and the supposed copy, and establish that there was an opportunity to copy. The copier who reproduces minor errors from the original (e.g. transposed digits in a table, or several spelling mistakes) is well on the road to trouble.

There are exceptions that allow limited copying without this being an infringement. Almost by their nature these cannot be defined exactly, but certain conventions are widely accepted. The main exceptions important to scientific and technical writers are:

- fair dealing for research or private study;
- fair dealing for criticism or review;
- certain exemptions for teaching or library use.

Under the first of these headings, single photocopies of a few pages of a book or a single paper from a journal for private study by the person making the photocopy is generally regarded as acceptable. Photocopying the library (or a friend's) copy of a whole book is not!

Fair dealing for criticism or review covers quoting, without specific permission, short passages (providing the source is acknowledged) to illustrate a critical point. Short quotations are generally regarded as isolated passages none of which exceed 400 words and a total of about 800 words from any one book. These limits should be reduced appreciably if quoting from shorter works such as research papers or articles.

In school textbooks a slightly more liberal use of quotations is allowed. In all cases sources must be acknowledged. When in doubt the safe rule is (and your publisher's contract will usually require you to do this) to seek from the copyright-holder or licensee of the relevant right, permission to quote any copyright material. This may be refused, in which case you must forget about using the material, but more commonly you will be granted permission to use the material subject to appropriate acknowledgement and sometimes only after payment of a fee. The fee may be only nominal, or it may be hefty. My advice is to pay a nominal fee of a few pounds or dollars for using, say, a table or a diagram, but consult your publisher if you are asked for £100 or $250 dollars to reproduce one diagram, or a few lines of text. Publishers may be able to help you get a better deal, they often carry more clout than an individual. Some publishers provide a draft letter for the use of their authors when writing for copyright permission. If you have to write your own letter be careful to indicate precisely what material you wish to use and what rights you are asking for. This may be governed by the markets for your book. You may only require rights to publish in certain countries, or you may require world rights.

Remember that permission is needed to reproduce complete tables, diagrams, graphs or photographs from copyright work unless there is an indication in that work that such permission is not needed. I mentioned in section 6.2 that some official government statistical services allow brief extracts from their published data without formal permission, providing the source is acknowledged. For example, the 1991 edition of the UK Central Statistics Office's *Social Trends* contains the statement:

Brief extracts of government departments' material from this publication may be reproduced, providing *Social Trends* is fully acknowledged as the source.

This is followed by an address to which application may be made for permission to reproduce larger extracts or to reproduce 'any other' material, i.e. any item which is not 'government departments' material'.

If you are writing a textbook you may want to use data or information in published work in your illustrative worked examples or as a basis for student exercises. Such use is usually regarded as being covered by the 'fair dealing' criteria for research or for criticism and review, or sometimes by the teaching exemption. You will, however, be in trouble if you copy a number of exercises from another textbook and include them in yours. Remember, too, that questions set by institutions and by examination bodies are copyright and (like a table or graph) each is a complete work in itself. Permission to use such questions must be sought from the appropriate examining body and a fee paid if necessary. In my experience no, or only a minimal, fee is charged for such use by most professional bodies; some have a requirement that the question must be reproduced exactly as printed in the examination paper without any modification to wording.

Infringement of copyright leaves the infringer open to a civil claim for damages. These will often exceed substantially the fee that might have been asked if you sought prior permission, especially if it can be shown that the breach of copyright led to loss for the holder. For example, if you used a large amount of material from a successful textbook and your book proved so popular (or was so much cheaper) that it virtually killed the market for the original text, you may expect a large claim for damages. In the USA in particular, commercial considerations play a major role in determining damage levels for infringement.

An unfortunate form of copyright infringement is book piracy, or the printing and sale of illegal editions or translations of books. These usually appear and are sold in countries that are either not parties to the international copyright conventions, or whose governments or law enforcement agencies are unhelpful about enforcing these agreements. Publishers are taking an increasingly tough line against book piracy. They have more resources than an individual author for fighting this evil. If you learn, or even suspect, that your book has been pirated, notify your publisher immediately.

A1.3 Moral rights

The UK and US copyright laws both recognize the so-called 'moral rights' of authors — sometimes called the rights of *paternity* and *integrity*. The right of paternity, which must be asserted by the author in writing when assigning copyright or when licensing rights, requires that the author is clearly identified in all versions of the work. Integrity is a right of protection against derogatory treatment of one's work, including any deletions, adaptations or alterations which might be prejudicial to the reputation of the author. This is one reason why publishers or their copy-editors should query with an author any proposed alteration of substance, for if it introduces an error or obscurity not in the author's original work, this could be an infringement of moral rights.

A further moral right is that a literary work may not be falsely attributed to you as the author. It is hard to visualize situations where this might occur in scientific and technical writing, although I suppose it is just conceivable that some unscrupulous company might produce an in-house or a publicity report and attribute it to an eminent academic who had done no more than make a few minor comments on a draft version.

A2 LIBEL

A2.1 Civil actions for damages

The laws of libel vary between countries and even between parts of countries. For example, there are separate, though not very different, libel laws for England and Scotland. Those in the USA differ again. However, broadly speaking, you may lay yourself open to civil claims for damages if you make a defamatory statement that can be proved to injure the reputation of an individual or individuals. It is rare for civil libel actions to rise from scientific or technical writing.

Nevertheless, some care should be exercised when, for example you criticize published work. If you have evidence that proves that a research worker used faked data in a published report, you will not be laying yourself open to libel if you publish your proof. You would be in potential trouble if you added the rider *This is not the first time this worker has used fictional data* unless you also had firm evidence to back that claim.

Publishers' editors and sub-editors are normally on the look-out for potentially libellous material and are likely to query any passage that they think might be libellous. Expressing an outrageous opinion may not in itself lay an author open to a civil libel action. For example, if you write that in your opinion all economists talk rubbish, many of your readers (especially those who are economists) may think you are talking rubbish yourself; but the statement is not libellous because it does not identify any particular economist. On the other hand, if you refer to a specific committee consisting of five economists and claim that one of them was asked to resign from his University because he embezzled departmental funds, you could be in trouble even if the statement were true and you did not name the individual. This is because, by your failure to name the individual, all five members of the committee may consider their reputation is damaged by implication.

One must also be careful of accidental libel. If J.W. Smith had admitted to using faked data in a paper published in 1957 you could be in trouble if you wrongly attributed this admission to another Smith, perhaps J.T. Smith, and the latter considered (as he probably would) that his reputation was damaged by your statement. This is a further reason for checking facts and taking care to get references correct.

A2.2 Criminal libel

Most countries have laws forbidding publication of certain types of material — typically writing that is obscene, seditious, incites racial hatred or is blasphemous. In some circumstances such writings may give rise to civil claims of the type considered in the last section, but they may also lead to criminal libel actions which could result in a fine or imprisonment rather than a straight (and sometimes hefty) award of damages.

In the UK, actions for criminal libel are less common than civil claims for damages, but prosecutions for obscenity may be brought. It is hard to visualize such cases arising with serious scientific and technical writing, though it is just possible that at the popular level some biologist might produce what is no more than a pornographic sex manual under the guise of a serious scientific treatment of sex. Here there is an analogue to that thin line between art and pornography.

A charge of sedition, or incitement to racial hatred, would seem to be a potential danger only in subjects like anthropology or the social sciences if controversial statements of a racist nature were made. It is unlikely that presenting evidence of, or fair comment about, intelligence differences between racial or ethnic groups would trigger charges of inciting racial hatred, but suggestions that lasting peace could only be attained by sterilizing all males of some named ethnic group might lead to such a charge.

Social scientists, especially political scientists, could run into trouble on matters of sedition. Reasoned criticism of parliament or the monarchy is acceptable providing it is not accompanied by advocating reform by unconstitutional or violent means.

British blasphemy laws are restricted in that they apply only to vilification of the Christian religion. It is now accepted that temperate and sober writing on religious topics, no matter how strong the anti-Christian bias, will not lead to prosecution, but if the impact of an attack on Christianity, the Bible or the Book of Common Prayer were adjudged to be such that they might offend Christians or lead to violent action (e.g. the burning of Churches), prosecution might follow. It would not be a defence to claim that the words were not intended to lead to violent action. Neither would that be a defence against incitement to racial hatred.

Again, for blasphemy, one does not visualize serious problems in this area for most scientific and technical writers, but if you are writing about the conflict between religion and science, or perhaps a treatise on comparative religion, make certain you do not get carried away. Remember also, that although the British law of blasphemy only covers Christianity, regrettable situations may arise from criticism of other faiths; the problem is highlighted in two words — Salman Rushdie.

A3 TAXATION

Nearly all countries (I know of only one exception) levy income tax on earnings from writing. Because you will often receive royalties or a fee for a media contribution gross (without deduction of tax), do not think there is no need to declare this income to the taxation authorities. In most countries a serious author will be taxed in the manner generally applicable to self-employed professional people and certain expenses may be deducted from gross income received. If your writing is little more than a hobby, your income is still liable to tax and you may have more difficulty claiming expenses than would be the case if you were a 'professional' author.

Because a large slice of the income from a highly successful book, which may take several years to write, is very often received over a much shorter period, many countries allow an author to 'spread' that income for taxation purposes and treat it as though it were earned over the several years during which the book was written. Under some tax regimes this will reduce considerably the tax liability that would otherwise arise. There may be special rules about royalties earned overseas. Sometimes you have to pay tax on these in the country of origin. Many countries have reciprocal taxation agreements to ensure that you are not 'double taxed' on such earnings.

The laws on taxation, and there are taxes other than income tax, are complex. In the UK advice is available to members from the Society of

Authors. In section 4.3.1 I advised that if you felt legal help necessary in connection with a publisher's contract, then you should choose a lawyer who specialized in matters to do with authorship. I give similar advice if your earnings from writing are such that they are likely to have important tax implications. Do not then seek the advice of any accountant, but choose one who specializes in matters relevant to authors. Remember though, that accountants, like lawyers, charge substantial fees for their services.

Appendix B

Preparing camera ready copy

B1 FORMAT FOR CRC CONFERENCE PROCEEDINGS

Camera ready copy (CRC) is widely used for published conference proceedings. It speeds publication to ensure topicality, but unfortunately lax supervision of some early ventures in this area led to shoddy end products, giving CRC a bad reputation. Not only have computer printer facilities improved, but publishers and editors of conference proceedings now usually take positive steps to ensure uniform and acceptable standards of CRC.

Because most contributed papers at a conference will have a different author or authors, all contributors need advanced and detailed instructions about how they are to submit CRC. The nature of these instructions will depend on several factors. If all papers are to be refereed prior to acceptance, it is common to ask for submissions to be in 'draft' CRC that complies as far as possible with instructions sent previously to all contributors. This gives the editor(s) an opportunity to point out any technical imperfections in layout. The editor(s) usually reserve a right to require contributors to make necessary amendments to the CRC format at the same time as they make any text changes resulting from the refereeing process. To allow for exceptional cases where, because they have not got appropriate technical facilities, authors may be unable to comply with all requests for format changes, there is a case for editor(s) providing back-up arrangements to have such revision carried out on behalf of authors who face technical problems. For example, if there is a specification that all work be set in *TeX* a few contributors may not have access to this package.

As I pointed out in section 10.2.2 contributors to conference proceedings are often sent special paper suitable for the photographic process; this often incorporates a blue-line margin. Authors are instructed to type within these margins to ensure a uniform text area on a page. This provision alone does not guarantee text will be submitted in a suitable format. To obtain reasonable uniformity between contributors, they will also need instructions on the following points, some of these are concerned with detailed format and others with *macrostyle:*

- the font to be used;
- point size for text and headings;
- spacing, positioning and style for title, author(s) names and affiliations, section and subsection headings;
- any requirements for abstracts, keywords, etc.;
- instructions about page numbering, running headers, etc.;
- rules for citing references;
- requirements for tables, diagrams etc., including format for headings and captions;
- format for, or restrictions on, photographs and any other art work.

It helps contributors if they are sent, with the instructions, three pages of sample material; this should consist of a 'first' page showing how to lay out titles and other headings; a typical text-page which might, if appropriate, include a section and subsection heading (showing how these are to be numbered and spaced relative to the main text), an example of a displayed formula and one or two citations; the third page should indicate how citations are to be set out in the reference list. The latter might look something like the examples given in section 2.2.5, p. 23.

Though detailed instructions will vary with circumstances, they should be such that most contributors can reasonably be expected to comply with them. For example, if you are an editor, specify fonts and point sizes that you are confident will be available to all, or nearly all contributors. Quality requirements must also be stated, e.g. laser or inkjet printer standard. Most modern computer/printer combinations provide a Times Roman font or a near equivalent proportionally-spaced scalable font. An editor who has doubts about the universal, or near universal, availability to contributors of this font might opt for the ubiquitous Courier font, because this would also cater for some contributors who have only a daisy-wheel printer. The price to be paid is that the book would have a typewritten, rather than a typeset, appearance.

Point size is important. Normally the text area per page in the printed book is reduced in the photographic process, so, in the original, the main text font size should be *at least* 12 point. If it is assumed that a laser or inkjet printer will be used scalable fonts should be available, and it may be wise to specify a 14 point size. This will probably scale to something like 11 point, but the scaling factor will depend on the size reduction. For simplicity it may be wise to use the same fonts for titles, headings, etc., highlighting them by using different point sizes and bold or italic, e.g. the title might be set in 24-point bold, the author(s) name(s) appearing in 15-point capitals and their affiliations in 14-point italic, all based on Times Roman. Section headings might be in 14-point capitals and subsection headings in 14-point bold, all in lower case except for the first letter and any essential capitals in each sub-section heading.

Spacing and positioning of titles, author(s) name(s), affiliations, section headings, must be clearly stated, e.g.'Title to commence 2 inches from top of page', etc. Instruction must be given about justification — usually either full justification (type flush with both left and right margins) or left justification only (i.e. ragged right margin). Rules about indentation at the start of paragraphs should be clear. It is customary **not** to indent the start of the first paragraph after a section or subsection heading, but to indent at the start of subsequent paragraphs. Features like this are easily demonstrated on specimen pages; these pages are also invaluable for illustrating the positioning and type of abstract (if any) that is required, the way keywords should be displayed and the way references are to be cited in text and how bibliographical details are to be set out in the *References* section. In particular, make it clear whether the Harvard or Vancouver systems should be used for references (section 2.5.5).

Instructions about tables and figures should give the required format for headings, captions etc., e.g. whether to use bold or italic in these, whether to spell out 'Figure' or use the abbreviation 'Fig.', and whether headings and captions should be in the same font and point size as the main text. A smaller point size is often used in tables. Again it is useful to give examples in the specimen pages sent to contributors.

Because page numbers in published proceedings cannot be determined until all contributions are received it is often best for the editor or printer to supply these, and sometimes also running headings for pages, at the printing stage. However, to avoid disasters, when submitting the CRC, folios should be numbered consecutively to indicate order. Editors should include an instruction as to how this is to be done. If a typing area has been indicated by blue margins the page numbers may sometimes be printed outwith these margins (because material outside the margins is not photographed), but the editor may prefer folio numbers to be lightly pencilled on the back of each folio.

Finally, no matter how good the instructions, there will be some contingency the editor forgot to cover! It is a good idea for editors to provide a contact name and address (with perhaps telephone, fax and e-mail information) that contributors with difficulties may contact, so that most problems can be resolved at an early stage. Problems I have met, even when the editor had sent good and clear instructions, include how to present the author/affiliation information when each of the several authors of a paper had different affiliations; how best to incorporate a glossary, where appendices or glossaries should be placed.

B2 CRC FOR BOOKS

Do not try to produce a book as CRC unless you or your publisher — preferably both — are enthusiastic about it. Unless you have an interest in typography, book design and typesetting, seek guidance on these matters from your publisher or some other expert before you start. Many publishers, including Spon, provide a guide for those preparing CRC. To do a good job you will need up-to-date word-processing or desktop publishing software and a modern printer — preferably laser with 600 dpi (dots per inch) resolution. Even that high resolution does not produce as sharp an impression as quality professional typesetting; a deficiency that is overcome by using a page text area and point size that is reduced by the camera to some 80 to 85% of the original size. The main text in this book is set in 14-point Times New Roman, but what you are reading now is approximately 11 point, because the size has been reduced to 80% of the original.

A common mistake made by amateur typesetters is to use too many fonts. Modern word-processor and printer combinations usually allow the use of some 20 fonts and an enthusiast can add dozens more. A good basic rule is not to use more than two fonts on a page. In this book, except for rare exceptions to illustrate some specific point I have used only one text font; Times New Roman for text and for section headings and its italic form for running headers at the top of a page. Chapter headings are in the bold version of the same font. I have used a different font for labels and text in graphs, but a common convention is to use the same font in these as is used in the main text. I have used a smaller size (12 point originally, which reduces to just under 10 point) in tables and in figure captions.

I have made more use of italic and bold than would be appropriate for most books. I explained why in section 3.2.6.

Modern word-processors allow one to add many embellishments, including a variety of borders and fancy edges to diagrams. These may look fine in advertising copy, but they detract in a book. Borders are often used for chapter headings, but use borders for text only when there is a clear advantage in doing so. I did it on p.120, where I wanted to illustrate the use of text 'in a box'. I also did it in Figure 8.1, where I wanted to give an impression of a letter typed on its own page and in that context a border identifies with the edge of a page, even though the dimensions may for technical reasons not be those of a typical page.

If you use two typefaces and Times Roman is your basic font, choose a contrasting type face such as Univers or Helvetica because these have a basic difference that immediately contrasts with Times New Roman. They are *sans serif* types, whereas Times is a *serif* type. *Serifs* are embellishments at the extremities of characters. The difference is readily seen in larger font sizes, e.g.

Times New Roman: **T**   Univers: **T**

Another typographic feature that may trap the unwary is line spacing. How much space should there be between a chapter heading and the first section heading? Between the end of a section and the next section heading? Between a section heading and the start of the text? Should these spacings be the same for subsections? How much space between text and a table heading? How does one use space within tables? How much space between the bottom of a figure and the caption, and between the caption and continuing text? Because these special spacings often involve half- or even quarter-line spacing, and the number of special spacings varies from page to page, the uncritical use of hard and fast rules may result in non-alignment of the last lines on opposite pages. This problem is often accentuated by a desire to avoid orphans and widows, i.e. the start of a new paragraph on the last line of a page or the ending of a paragraph as the first line on a new page. Further adjustments of line spacing may be needed to overcome these difficulties and to give a more pleasing page.

I illustrate some of these points by explaining how I dealt with them when writing Chapter 2 of this book. The text area on the printed page of this book is about 80 per cent of that on my CRC, which occupied an area of 232×147 mm. If you turn to the start of Chapter 2 on p.10 you will see the following:

- The page number does not appear on the first page of the chapter. It would look ugly and isolated at the top. Sometimes, when page numbers are put at the top (as part of the running header) on the remaining pages, the number is put at the bottom centre of the first page of each chapter. This is a matter of choice. I prefer not to do so.

- Some space is left at the top of the first page before the chapter number, which appears in 24-point bold in a shaded box. A line space (corresponding to one required for 24-point type) is left before the chapter title *Macrostyle*, which also appears in 24-point bold — in lower case apart from the first letter. There then follows an appreciable space before the first section heading which I set in 14-point Times New Roman capitals.

- There is no punctuation in chapter or section headings except for the decimal point in the section numbering system. There is a one-line space between the section heading 2.1 and the start of the text.

- The first paragraph in a section or subsection is not indented. Subsequent paragraphs have the first line indented (approximately 8 mm on the original CRC).

- Section 2.2 starts with a subsection 2.2.1. Subsection headings are in 14-point Times New Roman bold and all in lower case apart from the initial capital and any other essential capitals. There is a gradation in the spacing between the end of section 2.1 and the section header 2.2 (2 blank lines), between the section and sub-section heading, approximately 1 blank line, with an additional half-line blank between the subsection heading and the text. Juggling the exact values of such spacings is one way to line up the last line of text on a page with the last line of text on an opposite page when this is relevant. It is usually not relevant when the first page of a chapter is opposite the last page of the previous one, for the last page of a chapter is usually an incomplete page.

The remaining pages in Chapter 2 differ from the first page in having running headers. The font used is the italic version of 14-point Times. On odd numbered pages the running header is the title of the section (not the sub-section) on that page. It happens with this chapter that section 2.2 is very long and continues to the end of the chapter. In general, if a new section starts part-way down an odd-numbered page the header should be this new section heading, and not that of the section concluded on that page. On odd numbered (recto or right-hand) pages the page number appears at the top right. On even numbered (verso or left-hand) pages the chapter title — here *Macrostyle* — is used as the running header. The page number is at the top left for even numbered pages. There is slightly more than a one-line space between the running header and the first text line on each page.

On page 15 I included some material on conventions for section headings, together with examples. To avoid confusion with book section headings, which are all set flush left, I have put the illustrations indented some 60 mm, each displayed on a line of its own, using an extra half-line spacing to separate displays from the text. Again, such additional spacing may need adjusting (to something between one quarter and three quarters of a line of extra space) to line up the final line on the page with that on the opposite page.

Examples of tables occur on pages 17 and 18. In each, the table number is in bold and there is no full point at the end of the heading. Both headings and table content are in a smaller point size (12 point on the original CRC) than the body of the text. Spacing and horizontal rules have been used in a way that leaves the tables looking not too cluttered. Columns are clearly separated. It is often a matter of trial and error to achieve pleasant table layout. Tables should be placed as close as possible to relevant textual discussion. Some publishers like all tables to be at the top of a page (or else at the bottom) but there are often occasions when I prefer to break this rule to line up tables with text. I have done that on p.17. You will notice also on p.17 and p.18 that I have left quite a bit of space above and below Table 2.2 and Table 2.3. This improves appearance and helps to line up the final lines on the opposite pages 16 and 17. I preferred this solution to putting more text on p.17. This would have broken up the smooth flow of text around Table 2.3.

Figures 2.1 and 2.2 do not occupy complete pages, and I have put each at the top of a page — the preferred position for figures — although sometimes a figure may be 'centred' in a page. You will see that the text in the figures is in a sans serif type face (Arial) while the caption is in Times New Roman (12 point before reduction), the same font as the main text.

On p.14 there is an example of a device I use a lot in this book. Near the middle I highlight two points by displaying them indented on the left with a ● (known as a bullet) flush with the left text margin to draw attention to them. I mentioned in section 10.1 that several other symbols for bullets are commonly used, these include the square, ■, the diamond, ◆, pointing finger, ☛, arrow, ➜, and the triangle, ▸.

A final reminder: never forget that your final CRC, prepared usually immediately after copy-editor's comments have been taken into account, is the master copy for your book. Your final proof-reading must be done before this goes to the camera.

Appendix C

A glossary of publishing terms

This appendix explains some terms commonly used in publishing either by editors or by typesetters and, where appropriate, comments on points of specific relevance to scientific and technical writing.

Annotated bibliography *see* **Bibliography**.

Bibliography A list of publications relevant to the subject matter of a paper, article or book. Not all the entries in a bibliography need be cited in the text. Items are usually arranged alphabetically by author and give complete details including title, year of publication and the publisher for a book, or for a paper the title, name and volume number of the journal, and page numbers if relevant. Sometimes additional comments about the nature of a particular item are included; in this case the bibliography is described as an **annotated bibliography.** *See also* **References**.

Bold A heavy typeface used, for example, for headwords in this glossary. It is also used for emphasis or to highlight a term when it is first defined. In mathematics capital bold letters are used to denote matrices, e.g. **A, B, M,** while vectors are often indicated by lower case bold, e.g. **i, x, y**.

Bracket order In mathematics the standard order for brackets is [{()}], but note the special notation [x] for the integral part of x, i.e. [3.7]=3.

Contents In a book the **Contents** or **List of contents** is a list of parts (if any), chapter numbers and titles (and often section headings) placed in the **Prelims** at the front of a book, giving the first page number for each. It should not be confused with an **Index**, which is a key to topics. The index is put at the end of a book.

Display An item is displayed when it is printed, often centred or indented, on a separate line or lines broken from the main text.   Chemical and mathematical equations or formulae are often displayed. Quotations are sometimes displayed, especially if they are used to exemplify a point or are a focus for critical discussion. An example of a centred display is:

THIS LINE IS CENTRED

Often, as here, extra space is inserted between displayed and text lines. Double indentation is also used for displays, e.g.

> The text in this and the next line stand out from the body of the text because this paragraph is *double indented.*

While quotation marks are needed for a quote in text, these may be omitted if a quotation is displayed.

Drop capitals Drop capitals are sometimes used for the first text letter of a chapter. The practice used to be more common than it is today, although the effect is easily achieved with computer typesetting, e.g.

> This is an example of a drop capital. It is a device that is more widely used in fiction or art books or children's books. In the latter the capital is sometimes embellished with additional artwork.

em dash An 'em' dash or 'em' rule is approximately the width of the letter *m* in the point size currently in use. It is the common dash used — as here — to break up text in a sentence. The 'en' dash or 'en' rule is a slightly shorter dash used to break numerals, etc., e.g., pp. 27–33. Both the 'em' and 'en' dashes are longer than the hyphen (-). There is a tendency now to use either an 'en' dash or the even shorter -, which is strictly the minus sign, for either a minus or an 'em' or an 'en' dash without distinction.

em space A space between letters equivalent to that occupied by the letter *m* in the point size and font currently in use. With left justification only, the space between words is often kept uniform at the lesser 'en' space. If full justification (both left and right) is used, the space between words varies from line to line within certain tolerances that may be controlled by the typesetter.

en dash   *see* **em dash.**

en space   *see* **em space.**

Figures The word used to refer collectively to illustrations such as diagrams, graphs and line drawings; often (but not always) excluding photographs.

Each figure needs a caption and figures should be numbered with no distinction between different types of figure (e.g. graphs, diagrams and line drawings are all numbered in one sequence). Separate sequences may be used within each chapter, or the one sequence may be used throughout a book. In the former case use a decimal number system, the first number referring to the chapter. In captions, or in text references to figures, use consistently either the full spelling **Figure** or the abbreviation **Fig**.

Flush right The placing of text at the right end of a line. In headers in this book odd page numbers are placed flush right. The device is often used for identification numbers of formulae or equations that are displayed, e.g.

$$S(x) = a_0 + a_1x + a_2x^2 + \dots \qquad (7.15)$$

Folio A publisher's and printer's name for a manuscript page and also the typographical definition of a page number in proofs.

Font A typeface of a specified kind having defined characteristics. Sometimes each point-size of a given font is itself referred to as a separate font, but this usage is becoming less common with the widespread use of scalable fonts that allow virtually an infinity of point sizes. Some examples of different fonts are:

𝕭lackletter. 𝔍t would be tiring to read a book set in this font.

Clarendon Condensed Bold. Quite good for headings.

`Courier.  This is a common typewriter font.`

Script. Fine for wedding invitations.

Times New Roman. You may have seen this before.

Univers. Another one you may have seen before.

Footnote A note placed at the foot of a page to elucidate some point in the text on that page, e.g. a translation of a foreign passage, or a biographical note about a named person. Footnotes are usually printed in a different point size (and sometimes in a different font) to the main text. They are linked to the text item by a superscript symbol such as an asterisk or by a number. The latter is useful if there are several footnotes on one page. Although modern typesetting methods mean footnotes are no longer the typographer's nightmare that they were in the past, they generally look untidy and most publishers prefer authors not to use them.

Full justification   Text aligned to both left and right margins is said to be fully justified. Most publishers prefer manuscript text to be aligned with the left margin only, leaving the right end of lines 'ragged'. In printed books text is usually fully justified for prose, but not for poetry. Exceptions are sometimes made for certain blocks of text, e.g. text in narrow columns or in narrow boxes, where full justification may lead to ugly spacing between words or to excessive hyphenation.

Galleys   Galleys, or galley proofs, are unpaged proofs. Galleys are seldom supplied now because computer typesetting usually combines the process of typesetting and pagination. In the days of metal type-casting the galley was the tray in which the metal 'slugs' of cast type were stored prior to assembling into pages.

Glossary   A list of terms or concepts usually arranged in alphabetical order with a description of their meaning or usage.

Gullies   Spaces between sections of type or a subset of diagrams or photographs that may be assembled on one plate. Typically, gullies are the space between adjacent columns in a newspaper or magazine, or spaces that may be left between several related diagrams combined in one figure to show different phases of some process.

Hair space   A fine space between words or printed characters used largely in printing formulae, e.g. $\log x$ and $\sin x$ rather than $\log x$ and $\sin x$ or $\log x$ and $\sin x$.

Harvard system   A system for giving references in text by author(s) and date, e.g. *Smith and Jones (1944) showed that . . .* or *This result was proved some years ago (Gray and Brown, 1952).*

Headers *see* **Running headers**.

Headword   The leading word in each entry in a reference work such as a dictionary or encyclopedia, or in an ordered listing such as a glossary or index. Headwords are often printed in bold, as in this glossary.

Index   A reference list of topics covered and where to find details (usually specified by page numbers) placed at the end of a book. Some books include both an **Author index** and a **Subject index** . Do not confuse the latter with **Contents.**

Inferior *see* **Subscript**.

Italic A sloping type style sometimes used to stress importance, but bold is often preferred for this. I have sometimes put quoted passages, or extracts from quoted passages, or passages to illustrate a specific point in *italic* in this book. This avoids excessive use of quotation marks. In biology, generic names are conventionally written in *italic* (but not the corresponding popular names), e.g.

Primula vulgaris but Primrose.

In mathematics, variables should be printed in italic, but not constants or operators, e.g.

$y - 7$, dy/dx, $\log x$, a^x, (but e^x because 'e' is the exponential constant).

Justification   The alignment of text to a margin or about the line centre (central justification). 'Left justification' aligns text to the left margin, 'full justification' aligns to both left and right margins, while 'right justification' aligns to the right margin only and is less commonly used except to obtain special effects, as in the text box on p. 120 of this book. *See also* **Full justification**.

Kerning A device whereby the spacing between letters such as W and A is reduced to take account of the opposite adjacent slopes. For this letter combination the effect is to bring the letters closer together.

WALTER without kerning becomes
WALTER with kerning.

Landscape printing In landscape printing the text lines are parallel to the long side of the page as distinct from the more usual **portrait printing** (q.v.) where, as on this page, text lines run parallel to the shorter side of the page. Landscape format is sometimes appropriate for diagrams or for tables which have too many columns to fit a portrait format. A landscape page inserted in a book which is generally printed in the portrait format should be placed so that the book is rotated clockwise through ninety degrees to read the landscape text.

Leading   A term for adjustment of space between lines to allow for the use of special characters such as superscripts and subscripts where these might tend to merge with the normal line spacing. The term originated in the days of metal typesetting when additional lead was put between lines to provide this extra space. For example, without leading we have:

As a result of the corrosive action of H_2O and other substances in the pipeline a slow build-up of extra NH_4^+ will result.

Extra leading between the two lines makes it clear that the $+$ is a superscript to the NH_4 and not connected with the 2 in H_2O.

As a result of the corrosive action of H_2O and other substances in the pipeline a slow build-up of extra NH_4^+ will result.

Left justification *see* **Justification**

Line spacing   The space between each line of text. Single line spacing is customary in books; double (or 2-line) spacing is used in manuscripts. In printing, the spacing for headings and for separation of diagrams and tables from text usually calls for at least a single blank line to be inserted, and the space between the end of a section and section heading is usually 2 blank lines with 1 to 1.5 lines blank between the heading and the start of text in the new section. Line spacing is both point-size and font dependent. For 14 point Times Roman single line spacing is approximately 5.5 mm.

Lower case   Non-capital letters are referred to in printing terminology as lower case letters. Capitals are usually referred to either as 'Caps' or **upper case.**

Machine-readable codes Codes for typesetting instructions that are inserted in manuscripts on disc and which can be read and interpreted by appropriate typesetting software.

Manuscript The hard-copy presented by an author to a publisher. Despite the name 'manuscript' it is not hand-written. Minimum standards for manuscript are typewritten or dot matrix output of NLQ (near letter quality), but laser or inkjet printing is preferable.

Octavo   A commonly used page-size for books, having page dimensions 234×156 mm and the text area, including headers and footers (if any) is usually about 194×116 mm.

Orphans If the last line of text on a page is the first line of a paragraph this line is referred to as an orphan. It is regarded as poor typography to have an orphan. Adjustments in line spacing (often between headings and text or between tables and diagrams and text) are devices used to eliminate orphans.

Page proofs Proofs in which the text is divided into pages with appropriate running headers and page numbering. These are normally the final, and often the only, proofs sent to authors. Alterations to text in page proofs may well have an on-going effect that continues to the end of a chapter and in extreme cases to the end of a book. Such changes are costly and should be avoided if possible. Many publishers will charge authors for changes (other than those arising from printer's errors) made at this stage. The charge may run into hundreds of pounds or dollars.

Parentheses The correct name for what are often called 'round brackets' i.e. ( ). When using brackets within brackets it is usual to use square or curly brackets for the outer set and parentheses for the inner set, e.g. Professor Jones [the youngest professor in the University (born 1971)] took the chair. However, this construction is ugly and it would be better to replace the square brackets by commas or dashes.

Plates Commonly used to indicate high-quality photographic reproduction or artwork often printed on special paper. For technical reasons to do with economy in paper use when binding a book, plates are often grouped together. Plates are numbered in separate sequences from diagrams, often as Plate I, Plate II, Plate III, Plate IV, ...

Point size A measure of the height of type; 1 point = 1/72 of an inch. Thus the greater the point size, the larger the type. The point size of the print you are reading now is approximately 11 point; it has been scaled down from the original 14-point type used when preparing the CRC. In this book chapter headings are approximately 20 point (scaled down from 24 point).

Portrait printing In portrait printing the text lines are parallel to the shorter side of the page. This is the normal format for most books. *See also* **Landscape printing**.

Prelims The material at the front of a book before the introduction or first chapter, consisting usually of the half-title page, title page, list of contents, preface, foreword (if any) and sometimes acknowledgments or a glossary.

Proportional spacing Type in which the spacing between letters is adjusted according to the space required for each letter, e.g. the letter 'm' requires more space than the letter 'i'. Times Roman is a proportional spaced type. Courier is not proportionally spaced. The effect is indicated below in the spacing of 10 consecutive letters in each of these typefaces.

Times New Roman:
mmmmmmmmmm
iiiiiiiiii

```
Courier:
mmmmmmmmmmm
iiiiiiiii
```

Quotation marks The most common practice is to use single quotation marks, the opening mark being ' and the closing mark '. The corresponding double marks are " and ", but an alternative opening mark is ". If single quotation marks are used as standard, then double marks should be used for a quotation within a quotation, e.g.

> 'Then he said to me "You must not do that," taking me completely by surprise.'

Recto A right hand (odd numbered) page in a book. *See also* **Verso**.

References The bibliographic details of all works cited in the text as sources of information or to back up statements or claims made by an author. A list of references differs from a **bibliography** (q.v.), in that only items cited in the text are included in the former while the latter may also contain details of items not cited in the text and sometimes further explanatory matter.

Right justification *see* **Justification**.

Running headers or **headings** The headers at the top of each page which include the page number (unless this is given at the foot of a page). On even numbered, or **verso**, pages the running header is usually the title of the current chapter and on odd-numbered, or **recto,** pages it is the title of the current section. If the relevant titles are too long to fit comfortably on one line an abbreviated version is used, e.g. if a chapter title is *The role of dust particles in the atmosphere in the formation of fogs*, this might be abbreviated to *Role of dust particles in fogs* for the running header. There is no running header on the first page of a chapter.

Sans serif Type which has no embellishments at the extremes of characters. *See also* **Fonts** and **Serif**. This sentence is printed in a sans serif font.

Screening A technique used in printing to produce satisfactory shading in diagrams and graphs. Without the use of screens, shading on hand-drawn or computer generated graphs will sometimes be lost (or may become too pronounced) in the photographic phase of typesetting.

Serif Type which has embellishments at the extremes of characters. *See also* **Fonts** and **Sans serif**. The type in this sentence (Times New Roman) is a serif font.

Small capitals   Capitals of a smaller size than those in the point-size currently in use. These are often used in names in reference lists, etc., for all letters after the first in a name, e.g. PETER SPRENT, and in qualifications, e.g. BSc, PhD.

Square brackets   Brackets of the form [ ] commonly used in mathematical formulae and as brackets outside parentheses or rounded brackets. In mathematics the expression $[x]$ has the specific meaning 'the integral part of x'. *See also* **Bracket order** and **Parentheses**.

Subscript   Characters placed below the base line of text are referred to as subscripts, e.g., $n-1$ in a_{n-1}. Subscripts are sometimes called **inferiors**.

Superiors *see* **Superscripts**.

Superscript   Characters placed above the text line are referred to as superscripts, e.g. N^{15}. Superscripts are also called **Superiors**.

Table   A table is essentially a systematic presentation of data in two or more rows and columns. The data may be quantitative or qualitative. A distinction is usually made between a table and a list, the latter term being applied to items listed in a single column. Such lists should not be numbered as, or referred to as, tables.

Upper case   A term used by printers to describe capitals. *See also* **Lower case**.

Vancouver system   A system for giving references by consecutive (often superior) numbers. e.g. *It has been shown[1] that the previous hypothesis[2,3] is not correct.* Full details are given beside each number in a list of references at the end of a paper or article, or in a book at the end of each chapter or at the end of the book.

Verso   The left or even-numbered pages in a book. *See also* **Recto**.

Widows   When the last line of a paragraph is the first line of text on a page it is called a widow. Widows are usually avoided by adjusting line spacing, often at headings or between text and tables or diagrams.

References

Fisher, N.I. (1993) *Statistical Analysis of Circular Data.* Cambridge University Press, Cambridge.

Haddock, B.A. and Jones, C.W. (1977) Bacterial respiration. *Bacteriological Reviews*, **41**, 47–99.

Hinkelmann, K. (1993) Biometrics — our journal: some observations from the editor. *Biometric Bulletin*, **10** (4), 11–12.

Rahn, T. (1993) Research and development for railway systems in Germany. *Quarterly Report of the Rail Technical Research Institute*, **34**, 33–42.

Sprent, J.I. and Sprent, P. (1990) *Nitrogen Fixing Organisms — Pure and Applied Aspects.* Chapman & Hall, London.

Sprent, P. (1969) *Models in Regression and Related Topics.* Methuen & Co., London

Sprent, P. (1988) *Taking Risks — The Science of Uncertainty.* Penguin Books, London.

Swift, J. (1720) Letter to a young clergyman, 9 January 1720.

Further reading

Arrangement of the list

This reading list is in three sections; the first lists books that deal with general aspects of writing. Some, but by no means all, of those listed deal specifically with one or more aspects of scientific or technical writing. Section II lists general reference works for authors while Section III lists books relevant to specific topics dealt with in this book.

The comments after some entries draw attention to aspects likely to be of special interest.

There are many books that deal with particular types of writing and with marketing one's writing, but which do not cover scientific or technical writing in any detail. I have only listed a few of these, which I feel to be of special merit. If you are writing for the 'popular' end of the market you may find books of this type are useful both for their tips on style and their hints about marketing. You will find many general 'how-to-write' books on display in larger bookshops, especially in the USA where there is a high output of books of this type and perhaps a greater demand for them than there is in some other countries. Look at such books and read any that you feel may be helpful in the field in which you write.

Section I

Books about writing

In this section I list books in alphabetical order by author.

Barrass, R. (1978) *Scientists Must Write.* Chapman & Hall, London.
A classic for students and those who have to write as part of their job.

Blackwell, B. (1985) *Guide for Authors*. Basil Blackwell Publishers Ltd., Oxford.
Essentially a publisher's in-house guide, but contains many useful hints about approaching any publisher.

Blake, G. and Bly, R. (1993) *The Elements of Technical Writing*. Macmillan Publishing Company, New York.
Particularly useful if you are writing in, or for, the US market.

Butcher, J. (1993) *Copy-Editing*, 3rd edn. Cambridge University Press, Cambridge.
A UK standard reference for copy-editors; also useful for authors.

Day, R.A. (1989) *How to Write and Publish a Scientific Paper*, 3rd edn. Cambridge University Press, Cambridge.

Kirkman, J. (1992) *Good Style — Writing for science and technology*. E & FN Spon, London.
A treatise on style for scientific and technical writing from papers and reports to manuals, containing much that is more widely relevant.

Legat, M. (1989) *The Nuts and Bolts of Writing*. Robert Hale, London.
A highly readable introduction to the art of writing and how to approach publishers. Although it does not deal specifically with scientific and technical writing, nearly everything in this book is relevant, nevertheless, for authors in these fields.

O'Connor, M. (1991) *Writing Successfully in Science*. Chapman & Hall, London
An excellent guide for writers of scientific and technical papers as well as for writing theses, reviews and even a curriculum vitae.

Palmer, R. (1992) *Write in Style — A guide to good English*. E & FN Spon, London.
An entertaining down-to-earth manual about how to write better English.

Rubens, P.H. (1992) *Science and Technical Writing. A Manual of Style*. Henry Holt & Company, New York.

Shortland, M. and Gregory, J. (1991) *Communicating in Science: A Handbook*. Longman, London.

Turabian, K.L. (1987) *A Manual for Writers of Term Papers, Theses and Dissertations*, 5th edn. University of Chicago Press, Chicago.
The University of Chicago Press is a leading US publisher of guides for writers. This book, as the title implies, covers many aspects of scientific and technical writing.

Turk, C.C.R. and Kirkman, J. (1988) *Effective Writing — Improving scientific, technical and business communication,* 2nd edn. E & FN Spon, London.
Especially relevant to *have-to-do* scientific and technical writing.

Section II

Some useful reference works for writers

In this section I list books by title under one of several sub-headings. I list them this way because many of the works either have no named author or editor, or are better known by their title. Some are readily available in reference libraries and this is where many authors will use them.

Dictionaries

Dictionaries range from general language dictionaries to special subject dictionaries. I have not included special subject dictionaries in my list because most scientific or technical writers will be familiar with those relevant to their own discipline. When they need information on technical or scientific terms outwith their own discipline these are likely to be words in fairly wide use, and they can usually be found either in:

The New Shorter Oxford English Dictionary. Clarendon Press, Oxford.

or

Webster's Third New International Dictionary. Merriam-Webster Inc., Chicago.

The New Shorter Oxford English Dictionary was published in 1993. Its predecessor, *The Shorter Oxford English Dictionary* gave a less satisfactory coverage of scientific and technical terms. *Webster* not only gives a good coverage of these terms, but for many words where there are different American and English spellings, both are usually indicated.

Many authors will have library access to either or both of the above works, but a smaller dictionary will often suffice for everyday use. Most standard one-volume dictionaries produced by a publisher with a good reputation for reference works will prove adequate, although some are more comprehensive than others in their treatment of scientific and technical terms. Always use an up-to-date edition of any dictionary, for language, use and meaning of words are undergoing a continuing evolutionary change. My personal favourite for everyday use is:

The Concise Oxford Dictionary. Clarendon Press, Oxford.

Dictionaries that deal specifically with writing and publishing include:

The Oxford Dictionary for Writers and Editors. Clarendon Press, Oxford.

and the more specialist

The Oxford Dictionary for Scientific Writers and Editors. Clarendon Press, Oxford.

Although not strictly a dictionary, a closely related work that all authors should own is a good thesaurus, e.g.

Roget's Thesaurus of English Words and Phrases.
Peter Mark Roget published his Thesaurus in 1852 and the work is now long out of copyright. A number of publishers have produced revised, and themselves copyrighted, usually satisfactory, updated editions. As for a dictionary, you will be wise to use the latest, or at least a recent edition.

Reference books about language, grammar and style

A basic guide to the use of the English language is the work commonly referred to as *Fowler's Modern English Usage.* More formally titled:

A Dictionary of Modern English Usage by H.W. Fowler. 2nd edn revised by Sir Ernest Gowers, 1968. Clarendon Press, Oxford.

I have a slight preference for

The Complete Plain Words by Sir Ernest Gowers. Revised by S. Greenbaum and J. Whitcut, 1986. Her Majesty's Stationery Office, London.

The above works are concerned with the grammatical structure of prose. Points of style at the author, publisher and printer interface are fully covered in *Hart's Rules.* More formally titled:

Hart's Rules for Compositors and Readers at the University Press Oxford. 39th edn., 1983. Oxford University Press, Oxford.

Two general guides to style that I have often found useful, especially when writing for the popular market, are:

The Times Guide to English Style and Usage. Ed. Simon Jenkins. Times Books, London.
The Economist Pocket Style Book. The Economist Publications Limited, London.

When writing for or in the US market a useful guide is:

The US Government Printing Office Style Manual. US Government Printing Office, Washington.

Authors' guides to marketing

Important guides for writers that contain invaluable information about publishers and about the press and magazines as well as general advice on writing are the four annuals listed below. Always use the latest available editions:

Writers' and Artists' Yearbook. A & C Black, London.
The Writer's Handbook. Macmillan, London.
The Writers' Market. Writer's Digest Books, Cincinnati.
The Writer's Handbook. The Writer Inc., Boston.

Section III

Books about specific aspects of writing

Publishing

Legat, M. (1991) *An Author's Guide to Publishing,* revised edn. Robert Hale, London.
   An excellent account of the author and publisher interface by a publisher-turned-author.

Finch, P. (1987) *How to Publish Yourself.* Allison & Busby, London.
Mullholland, H. (1984) *Guide to Self-publishing.* Harry Mullholland, Wirral.
   Two good books on self-publishing from users of that route into print.

Specific topics

Anon. (1989) *University of Minnesota Style Manual.* University of Minnesota, Minneapolis.
   This is a typical example of a style book issued by a large organization primarily for internal use. Many such manuals, like this one, contain a substantial body of information useful to all writers.

Clark, M. (1992) *A Plain TeX Primer*. Oxford University Press, Oxford.
A useful guide for those using the *TeX* program for preparation of scientific or technical manuscripts.

Cooper, B.M. (1990) *Writing Technical Reports*. Penguin Books, London.

Ehrenberg, A.S.C. (1982) *A Primer in Data Reduction* . John Wiley & Sons. Chichester.

Hedge, M.N. and Davis, D.J. (1992). *A Singular Manual of Textbook Preparation*. Singular Publishing Group, San Diego.

Higham, N.J. (1993) *A Handbook of Writing for the Mathematical Sciences*. Society for Industrial and Applied Mathematics. Philadelphia.

Horstmann, R. (1991) Writing for Radio, 2nd edn. A & C Black, London.
An experienced broadcaster's guide to all aspects of writing for radio.

Longmate, N. Writing for the BBC. British Broadcasting Corporation, London.
The British Broadcasting Corporation's own guide to writing for radio and television. It is frequently updated. Always use the latest edition as requirements change rapidly.

Milne, P. (1991) *Presentation Graphics for Engineering Science and Business*. Chapman & Hall, London.
A guide to practical graphics generated by computer.

Tufte, E.R. (1983) *The Visual Display of Quantitative Information*. Graphics Press, Connecticut.
A must for the serious user of graphs.

Popular Science

Gardner, M. (1985) *Mathematical Magic Show*. Penguin Books, London.
Gleick. J. (1987) *Chaos —Making a New Science*. Cardinal/Penguin Books, New York/London.
Hawking, S. (1988) *A Brief History of Time* Bantam Press, London.
Huff, D. (1954). *How to Lie with Statistics*. Victor Gollancz/ Penguin Books, London.
Pearce, F. (1987) *Acid Rain*. Penguin Books, London
Penrose, R. (1989) *The Emperor's New Mind*. Oxford University Press, Oxford. Vintage Books, London.
These six books provide contrasting examples of successful books about science written for the layman or for scientists who are not expert in the field. That by Martin Gardner is one of many by him that have been published both in the USA

and the UK. I admit to a bias in selecting books for this list, my choice being influenced by topics that appeal to me.

Copyright

Cavendish, M. and Pool, K. (1993) *Handbook on Copyright in British Publishing Practice,* 3rd edn, Cassell & Co., London.

Dworkin, G. and Taylor, R. (1989) *Blackstone's Guide to the Copyright, Designs and Patents Act, 1988.* Blackstone Press, London.

Index